NEW RESEARCH ON ANGIOGENESIS INHIBITORS

NEW RESEARCH ON ANGIOGENESIS INHIBITORS

ELMER T. SKINARD
EDITOR

Nova Biomedical Books
New York

For permission to use material from this book please contact us:
Telephone 631-231-7269; Fax 631-231-8175
Web Site: http://www.novapublishers.com

LIBRARY OF CONGRESS CATALOGING-IN-PUBLICATION DATA

New research on angiogenesis inhibitors / Elmer T. Skinard (editor).
p. cm.
Includes bibliographical and index.
ISBN 1-60021-473-8
1. Neovascularization inhibitors. 2. Blood-vessels—Growth. 3. Growth factors. I.Skinard, Elmer T.
[DNLM: 1. Angiogenesis Inhibitors—therapeutic use. 2. Neoplasms—drug therapy.
3. Neovascularization, Pathologic. QZ 267 N5328 2006]
QP106.6.N494 2006
612.1'3--dc22

2006031271

Published by Nova Science Publishers, Inc. ✦ New York

Contents

Preface		**vii**
Chapter I	Anti-Angiogenic Strategies in Hepatocellular Carcinoma: Current Status *D. Zacharoulis, C. Hatzitheofilou, E. Athanasiou and S. Zacharoulis*	**1**
Chapter II	Is there a Role for Angiogenesis Inhibition in Rheumatoid Arthritis? *Mohammed Ali Akhavani and Ewa Paleolog*	**23**
Chapter III	Antiangiogenic Property of Metronomic Chemotherapy *Katsumasa Kuroi and Masakazu Toi*	**55**
Chapter IV	Combined Radioimmunotherapy and Antiangiogenic Therapy to Treat Cancer: Validation Studies in Animal Models *Seigo Kinuya*	**77**
Chapter V	Quinones as Antiangiogenic Agents *Manal Chatila and Hala Gali-Muhtasib*	**99**
Chapter VI	Bevacizumab (Avastin) in Neovascular Ocular Disease *Nathan J. Rudometkin, Gian Paolo Giuliari and Antonio P Ciardella*	**115**
Chapter VII	Angiogenesis Inhibitors and Diabetic Retinopathy *Sho-ichi Yamagishi, Kazuo Nakamura, Takanori Matsui, and Takafumi Yoshida*	**135**
Index		**147**

PREFACE

Angiogenesis — the growth of new blood vessels — is an important natural process occurring in the body, both for health and as related to disease. Angiogenesis occurs in the healthy body to help heal wounds and to help restore blood flow to tissues after injury or insult. In females, angiogenesis also occurs during the monthly reproductive cycle (to rebuild the uterus lining, to mature the egg during ovulation) and during pregnancy (to build the placenta, the circulation between mother and fetus). The healthy body controls angiogenesis through a series of "on" and "off" switches. The main "on" switches are known as angiogenesis-stimulating growth factors. The main "off switches" are known as angiogenesis inhibitors. When angiogenic growth factors are produced in excess of angiogenesis inhibitors the balance is tipped in favor of blood vessel growth. When inhibitors are present in excess of stimulators, angiogenesis is stopped. The normal, healthy body maintains a perfect balance of angiogenesis modulators. In general, angiogenesis is "turned off" by the production of more inhibitors than stimulators. Tumor angiogenesis is the proliferation of a network of blood vessels that penetrates into cancerous growths, supplying nutrients and oxygen and removing waste products. Tumor angiogenesis actually starts with cancerous tumor cells releasing molecules that send signals to surrounding normal host tissue. This signaling activates certain genes in the host tissue that, in turn, cause proteins to encourage growth of new blood vessels. This new book examines its angiogenesis within the context of theory and its applications to cancer treatment.

Chapter I - Hepatocellular carcinoma is the third leading cause of cancer deaths and the fifth most common cancer in the world, accounting for over 1 million cases annually. In adults it remains a lethal disease and is identified clinically at an advanced stage [1-3]. Primary neoplasms of the liver represent 0.5 to 2% of malignancies in children. Eighty percent of malignant primary liver tumors in children are hepatoblastomas (HB), and 20% are hepatocellular carcinomas (HCC) [4].

Surgical resection has been considered the optimal treatment approach for HCC, but only a small proportion of patients qualify for surgery and there is a high rate of recurrence. Liver transplantation has been successful in treating limited stage HCC, affecting cure of both the tumor and underlying cirrhosis. However only a minority of patients qualify for transplantation. Chemoembolization has been shown to prolong survival in selected patients who do not qualify for transplantation or resection. Other local therapies, such as cryotherapy, radiofrequency ablation, have been shown to be effective in reducing the tumor size without prolonging necessarily the survival. Standard chemotherapy is poorly tolerated and overall response rates are low.

Angiogenesis, (the formation of new blood vessels) is an essential element in cancer biology. The progress that has been achieved over the past few decades in understanding the molecular pathophysiology of tumor angiogenesis is enough so that data can be reported in a more tumor- specific manner. Innovative approaches such as targeting the non-transformed, less resistant, tumor supporting endothelial cells are currently under investigation in HCC. Since a lot of interesting data have been reported specifically in HCC angiogenesis with immediate application into therapeutics, this review will focus on the current knowledge of the known factors that participate in HCC angiogenesis and the relevant clinical implications.

Chapter II - The last decade has heralded the advent of 'biological therapies'. These treatments are specifically aimed at biologically relevant targets, rather than relieving signs and symptoms of disease, or generally (and possibly non-selectively) modifying the disease process. In general, such biological targets have been identified using a tripartite approach, involving detection of the target molecule in a given disease, identification of a clear biological and functional role for the target, and finally demonstration of the efficacy of a therapy targeting the molecule in a relevant disease system. The fundamental model of a biological therapy is inhibition of tumour necrosis factor (TNF) α. The seminal studies demonstrating that TNFα is expressed in rheumatoid arthritis (RA) and that blockade of TNFα reduced disease severity in an *in vivo* model eventually resulted in the approval of anti-TNFα for the treatment of RA.

In oncology, biological therapies frequently target the vasculature, which fulfils a gatekeeper role in cell recruitment, through expression of leukocyte activators and chemoattractants, as well as adhesion molecules. Endothelial cells also play a role in the growth of new blood vessels, termed 'angiogenesis', which is vital for efficient supply of oxygen and nutrients to tissue. Angiogenesis has been a putative target in cancer since it was first linked to tumour growth and metastases in the 1970s. The most successful anti-angiogenic approach is antibody targeting the angiogenic stimulus vascular endothelial growth factor (VEGF). This antibody, bevacizumab, was approved in combination with chemotherapy for patients with colorectal cancer.

However, biological therapies targeting angiogenesis may have a broader application in the clinic. In RA, the expansion of the synovial tissue necessitates an alteration in the synovial vascular supply, to cope with the increased requirement for oxygen and nutrients. VEGF and many other pro-angiogenic factors have been demonstrated to be expressed in RA. Several studies have shown that targeting angiogenesis in animal models of arthritis ameliorates disease. It is therefore likely that, as in the oncological setting, angiogenesis is fundamental in RA and represents a potentially attractive target for therapy.

While the impact of TNFα inhibition on management of RA cannot be understated, the incidence of infections, and the lack of responsiveness in certain patients, supports the need for further therapeutic options. The focus of this chapter is to discuss the role of the vasculature in RA, particularly in the light of observations using therapies such as anti-TNFα, and on the potential for development of vascular-targeted therapies for treatment of RA.

Furthermore, the potential of approaches targeting VEGF, or other angiogenic factors, will be discussed.

Chapter III - Standard chemotherapy regimens are usually designed to kill as many tumor cells as possible by treating patients with the maximum tolerated dose (MTD) of cytotoxic agents. These agents are administered in a pulsed manner with breaks between cycles to allow recovery of normal tissues, thus limiting the toxicity of these agents. This strategy has led to a profound increase in the survival of children and adults with certain types of cancer. However, for the majority of solid tumors, particularly those at an advanced stage or associated with metastatic disease, neither complete tumor eradication nor sustained regression is often achieved. Moreover, MTD-based regimens are often associated with unpleasant or serious adverse effects that compromise the quality of life of the patient. The MTD concept targets the tumor cells themselves, and does not take into account the importance of the organization of a tumor, which consists of tumor cells and stromal tissue including endothelial cells (ECs).

In contrast, recent evidence suggests that most chemotherapeutic agents are capable of inhibiting tumor angiogenesis by targeting tumor ECs, and that metronomic, frequent or continuous administration of the same agent at low doses (a tenth to a third of the maximum tolerated dose) might optimize its antiangiogenic properties. In fact, the results of preclinical and clinical studies have been encouraging, especially when drugs are used in combination with new molecularly targeted agents such as specific angiogenesis inhibitors. By targeting normal, terminally differentiated and genetically stable ECs, the risk of acquired drug resistance might be reduced. Through the use of metronomic chemotherapy, it may be possible to re-evaluate not only the re-treatment of refractory cancers with agents that have previously failed, but also agents that have traditionally been considered inactive or ineffective against a particular type of cancer. Moreover, as some of the most useful agents tested in both *in vitro* and *in vivo* models include vinca alkaloids, taxanes, and antimetabolites, the principle of this concept does not necessarily require the development of so-called "magic bullets".

At present, however, it remains difficult to determine the optimal doses and schedules for metronomic chemotherapy, as a rapid objective response is absent, and identification of MTD using conventional toxicity criteria is relatively straightforward. Also, it is still unclear whether metronomic chemotherapy actually exerts an antiangiogenic effect that contributes to putative antitumor efficacy, and there is currently no standard for monitoring and quantifying the angiogenic response during therapy. Nevertheless, from a clinical view point, metronomic chemotherapy appears promising, not only because it reduces toxicity and the need for growth-factor support or anti-emetic drugs, but also because it may improve antitumor effects. The actual significance of this concept should be confirmed by much larger, prospective, and randomized controlled trials that compare metronomic dosing chemotherapy schedules with conventional ones. In addition, novel functional or angiogenesis-specific markers and new high-resolution techniques for blood flow imaging and analysis are essential for monitoring of angiogenesis.

Chapter IV - Radioimmunotherapy (RIT) employing radiolabeled monoclonal antibodies (MAb) against tumors has been extensively tested for more than two decades. However, in most cases, the results have demonstrated limited success in bulky disease. In contrast, several reports suggest the role of RIT as an adjuvant treatment of cancer. Neovascularization is a prerequisite for the growth of solid tumors; therefore, antiangiogenic therapy employing agents that suppress endothelial cell growth has been investigated. Antiangiogenic therapy does not generally cause bone marrow suppression or gastrointestinal symptoms. However, antiangiogenic therapy is inhibitory or cytostatic only, which may maintain lesions at limited size, a situation referred to as tumor dormancy. Therefore, combination with cytotoxic treatment is required for the extermination of tumor cells. We hypothesize that the combined utility of antiangiogenic therapy and RIT may cooperatively effect treatment of cancer lesions. The rationale for this hypothesis is based on suppression of tumor endothelial compartments by antiangiogenic agents as well as the direct action of RIT against tumor cells. Additionally, this combination may not increase bone marrow toxicity, which is the major determinant of maximal tolerated dose of RIT. This strategy was examined in a subcutaneous xenograft model and a liver metastasis model of human colon cancer cells using ^{131}I-labeled anti-colorectal monoclonal antibody and two agents that are known to exhibit antiangiogenic property, 2-methoxyestradiol and thalidomide.

Chapter V - Quinones are widely distributed among all respiring organisms. They participate in important reactions including photosynthesis, respiration and blood coagulation. Quinones have been used as antimalarial, antibacterial and anticancer agents. In addition to their redox-cycling capacity, they have been found to influence various signaling pathways known to affect cell proliferation, apoptosis and migration. Some quinones have been shown to decrease the angiogenic response, which is a prerequisite for tumor growth and metastasis. In this review, we attempt to describe the antitumor and antiangiogenic potential of benzoquinone ansamysins, anthracyclines, naphthoquinones, in addition to other interesting quinones.

Chapter VI - Angiogenesis is currently one of the most important topics in medicine. The regulation and control of blood vessel growth is important for both normal and diseased tissues. Vascular endothelial growth factor (VEGF) with its various subtypes is a vital signal in the pathway of vascular proliferation. Expression of VEGF has been noted in multiple types of cancer and in ocular diseases with vascular proliferation. Bevacizumab (Avastin) is a potent monoclonal antibody against all isoforms of VEGF that has been successfully used to treat colorectal, breast and lung cancer. The success of a related medicine, ranibizumab for treating neovascularization age related macular degeneration (AMD), prompted investigators to use bevacizumab "off-label" for neovascular AMD. Promising results have been found in treating exudative AMD both intravenously and more commonly through direct intravitreal injection with impressive improvements in patient's vision, ocular coherence tomography (OCT) and angiographic findings. Bevacizumab in both animal and human subjects appears to be safe and well tolerated in short-term evaluation. This antibody has been shown to penetrate extensively throughout the retina in albino rabbits. Other small series report successful treatment of proliferative diabetic retinopathy and retinal venous occlusive disease.

Over seven thousand intravitreal injections of bevacizumab have been administered in twelve countries as reported in the Avastin Safety Survey. To this point bevacizumab appears to be a safe and efficacious treatment of neovascular AMD and possibly other retinal vascular diseases. Despite the reported short-term success large controlled trials are necessary to clearly define the optimal treatment protocols and the precise frequency of adverse events. Consequently, care must be taken when treating patients with bevacizumab with detailed discussions about the possible adverse events of this "off-label" treatment. Currently, intravitreal bevacizumab appears to be a promising and cost-effective treatment for a variety of important ocular diseases.

Chapter VII - Diabetic retinopathy is one of the miserable microvascular complications in diabetes and is a leading cause of acquired blindness among the people of occupational age. Chronic hyperglycemia is a major initiator of diabetic retinopathy. Two recent large prospective clinical studies have shown that intensive blood glucose control reduces microvascular complications among patients with diabetes. However, strict control of hyperglycemia is often difficult to maintain and may increase the risk of severe hypoglycemia in diabetic patients. Furthermore, therapeutic options for the treatment of proliferative diabetic retinopathy, photocoagulation and vitrectomy, are limited by considerable side effects. Therefore, to develop novel therapeutic strategies that specifically target diabetic retinopathy is desired for patients with diabetes. This article summarizes the current understanding of anti-angiogenic therapy for diabetic retinopathy, especially focusing on pigment epithelium-derived factor (PEDF).

In: New Research on Angiogenesis Inhibitors
Editor: Elmer T. Skinard, pp. 1-21

ISBN: 1-60021-473-8

Chapter I

ANTI-ANGIOGENIC STRATEGIES IN HEPATOCELLULAR CARCINOMA: CURRENT STATUS

D. Zacharoulis[1], C. Hatzitheofilou[1], E. Athanasiou[1] and S. Zacharoulis[2]
[1]Department of Surgery, University of Thessally, Larisa, Greece;
[2]Department of Pediatric Hematology/Oncology, University of Southern California, Los Angeles, CA, USA

1. INTRODUCTION

Hepatocellular carcinoma is the third leading cause of cancer deaths and the fifth most common cancer in the world, accounting for over 1 million cases annually. In adults it remains a lethal disease and is identified clinically at an advanced stage [1-3]. Primary neoplasms of the liver represent 0.5 to 2% of malignancies in children. Eighty percent of malignant primary liver tumors in children are hepatoblastomas (HB), and 20% are hepatocellular carcinomas (HCC) [4].

Surgical resection has been considered the optimal treatment approach for HCC, but only a small proportion of patients qualify for surgery and there is a high rate of recurrence. Liver transplantation has been successful in treating limited stage HCC, affecting cure of both the tumor and underlying cirrhosis. However only a minority of patients qualify for transplantation. Chemoembolization has been shown to prolong survival in selected patients who do not qualify for transplantation or resection. Other local therapies, such as cryotherapy, radiofrequency ablation, have been shown to be effective in reducing the tumor size without prolonging necessarily the survival. Standard chemotherapy is poorly tolerated and overall response rates are low.

Angiogenesis, (the formation of new blood vessels) is an essential element in cancer biology. The progress that has been achieved over the past few decades in understanding the molecular pathophysiology of tumor angiogenesis is enough so that data can be reported in a more tumor- specific manner. Innovative approaches such as targeting the non-transformed, less resistant, tumor supporting endothelial cells are currently under investigation in HCC. Since a lot of interesting data have been reported specifically in HCC angiogenesis with immediate application into therapeutics, this review will focus on the current knowledge of the known factors that participate in HCC angiogenesis and the relevant clinical implications.

2. Tumor Angiogenesis

2.1. Basic Principles

It has been very well documented that tumors in order to grow beyond 10mm^3 require the formation of new blood vessels [5]. The advances that have been made over the past decades in the field of tumor angiogenesis lead to the realization that it is an oversimplification to consider the role of tumor blood vasculature is restricted to the "delivery of nutrients to the tumor cells". Instead, the tumor associated endothelial cells create complex autocrine and paracrine loops between the tumor and its microenvironment.

The onset of neovascularization in primary tumors can be relatively sudden, described as the "angiogenic switch". It is widely accepted that the "angiogenic switch" is "off" when the effect of pro-angiogenic molecules is balanced by that of anti-angiogenic molecules, and is "on" when the net balance is tipped in favor of angiogenesis. When a primary tumor first arises, proliferation of cancer cells may be balanced by apoptosis and the tumor may remain undetectable for years. This tumor dormancy period can be long until the *in situ* stage, when neovascularization occurs [6,7].

The recruitment of new blood vessels by a tumor is a multistep process and the factors involved in each of these steps have been the subject of intense investigation.The initial step involves degradation of the underlying basement membrane by proteinases such as matrix metalloproteinases (MMPs) [8,9], particularly MMP-2 and MMP-9 and plasminogen activators. Vessels are further destabilized by angiopoietin -2 (Ang-2), a ligand for the Tie-2 receptor. Endothelial cells then migrate, proliferate and assemble into tubes [10]. These cells originate not only from the pre-existing vessel but also are derived from a population of circulating endothelial precursor cells [11]. Numerous molecules stimulate endothelial proliferation, migration and assembly, including the vascular endothelial growth factor (VEGF), angiopoietin-1 (Ang-1), and basic fibroblastic growth factor (bFGF). Cell matrix receptors, such as αvβ3 and αvβ5 integrin mediate cell spreading and migration [12]. Maturation of nascent vessels involves formation of a new basement membrane and investment of new vessels with pericytes and smooth muscle cells. Recruitment of the smooth muscle cells is mediated through platelet derived growth factor (PDGF-BB) [13] whereas signaling by TGF-β1 and Ang1/Tie2 stabilizes the interaction between endothelial and smooth muscle cells. Survival factors of the new endothelium include VEGF (through interaction with VE-cadherin) and Ang -1 [14].

Below the current knowledge of the most important angiogenic factors in HCC biology are reviewed (figure 1).

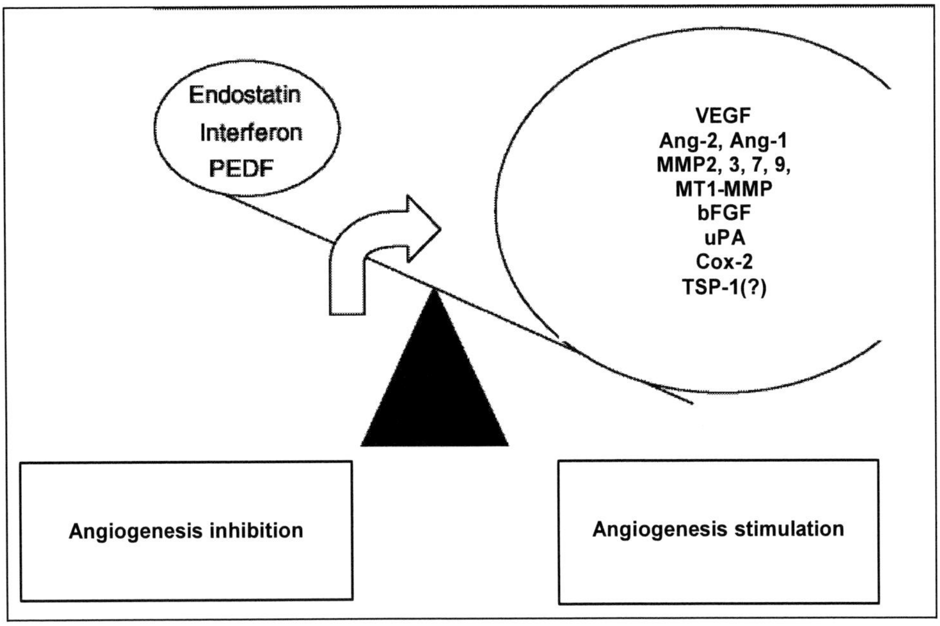

Figure 1. Angiogenesis stimulators and inhibitors in HCC. Ang: Angiopoietin; bEGF: Basic fibroblastic growth factor; Cox: Cyclooxygenase; HCC: Hepatocellular carcinoma; MMP: Matrix metalloproteinase; PEDF: Pigment

2.2. Angiogenic Factors

2.2.1. VEGF and its Receptors

The vascular endothelial factor (VEGF) is one of the most potent angiogenic factor. Yoshiji *et al* [15] showed by delivering the VEGF gene within the retroviral vector that overexpression of VEGF was associated with marked increase in HCC tumor development accompanied by augmentation of neovascularization. VEGF induces disruption of tight junctions and may promote HCC invasion into normal liver parenchyma [16]. Several studies have demonstrated an increased VEGF expression in tumor hepatocytes [17-19]. VEGF has also been found less intensively expressed in the higher-grade HCC component, not expressed in sarcomatous areas, while VEGF is expressed in the extracellular matrix [18], supporting the hypothesis of paracrine activation of VEGF at the level of tumor stroma. Overexpression of VEGF has been found in HCC compared with cirrhosis or normal liver and is strongly increased in areas directly adjacent to necrotic/hypoxic regions indicating that hypoxia is a central stimulus of angiogenesis in human HCC [19].

Serum VEGF level in hepatocellular carcinoma patients showed a positive correlation with tumor size and metastasis [20,21]. Furthermore when serum VEGF levels were corrected for platelet counts, they were found to be higher in patients with advanced stage and portal vein thrombosis [22] and associated with poorer outcome after resection [17].

The main receptors that seem to be involved in initiating signal transduction cascades in reponse to VEGF comprise a family of closely related receptor tyrosine kinases consisting of three members termed VEGFR-1 (known as Flt-1) , VEGFR-2 (known also as KDR or flk1

), and VEGFR-3 (previously known as Flt- -3). The expression of the VEGFR's in HCC has been well studied. In non-cancerous area, VEGFR-1 and VEGFR-2 are mainly found in macrophages including Kupffer cells in vascular endothelial cells in the portal veins and arteries within portal tracts, and VEGFR-2 is also found in some sinusoidal endothelial cells [23]. In cancerous area, VEGFR-1 and VEGFR-2 are found in some macrophages, as well as in the endothelial cells of intratumoral blood vessels [18].

2.2.2. Angiopoietins

Similar to VEGF, angiopoietins have important roles in angiogenesis. All of the known angiopoietins bind primarily to Tie -2 and is unclear whether there are independent ligands for the second Tie receptor, Tie-1,or whether the known angiopoietins, under some circumstances also engage to Tie-1, perhaps as a second component in an heterodimerized complex [24].

Unlike mouse lacking VEGF or VEGFR-2, embryos lacking Ang1 or Tie-2, develop rather a normal, primary vasculature. More precisely, endothelial cells in these embryos fail to associate appropriately with underlying support cells, which are the cells that provide the Ang1 that acts on endothelial Tie2 receptors [25]. Instead of supplying an instructive signal, that directs vascular remodeling, it is hypothesized that Ang1 has more a permissive role, by optimizing the manner by which endothelial cells integrate with supporting cells, thus allowing them to receive other critical signals from their environment.

Angipoietin -2 overexpression in the embryonic endothelium, resulted in embryonic death due to defects those of Ang1 or Tie 2 knockouts , demonstrating that Ang- 2 could act as Tie -2 antagonist under some circumstances [26]. The functional role of Ang -2 in angiogenesis is not very clear. It is hypothesized that Ang-2 may provide a key de-stabilizing signal, involved in initiating angiogenic remodeling [24].

Increased expression of Angiopoietin-1 and Angiopoietin-2 mRNAs has been detected in HCC [14,27-34]. Angiopoietin-1 and Angiopoietin-2 (at the mRNA level, as well as immunohistochemically) has been detected in hepatoma cells, hepatic stellate cells, and smooth muscle cells, whereas Tie-2 was detected in endothelial cells, hepatic stellate cells and smooth muscle cells [27]. Increased expression of Angiopoietin-2 and Angiopoietin-2 mRNA is associated with tumor dedifferentiation. The expression of Angiopoietin-1 and Angiopoietin-2 correlated with HCC vascularity [29,33,34]. The up-regulation of the angiopoietins unlike VEGF might mot mediated through hypoxia in HCC [32].

2.2.3. Matrix Metalloproteinases (MMPs)

One essential component of the angiogenic process is the degradation of extracellular matrix and basement membrane. This process is mediated by several proteases, including the MMPs. There are at least 20 different MMPs known in humans with distinct substrate specificities [35]. Moreover, MMPs promote the release of extracellular matrix-bound or cell surface bound cytokines [36] such as VEGF, which then can regulate also angiogenesis. Additionally MMPs contribute to angiogenesis by facilitating the release of bone marrow derived endothelial cells that participate in tumor vasculature, by cleaving *kit* ligand [11].

Three main networks of MMPs are predominantly known to be involved in the case of HCC, that is, MMP-2 and MT1-MMP in the carcinogenesis and progression, MMP-7 and

MMP-9 in the capsular infiltration and portal vein invasion, as well as MMP-3 and MMP-7 in the progression of HCC. Furthermore, MT1-MMP appeared to be the most important factor of the family in HCC because of its widespread pattern of expression [37].

2.2.4. Basic Fibroblastic Growth Factor (bFGF)

Basic fibroblastic growth factor is another potent angiogenesis stimulator, involved in endothelial cell proliferation, migration, sprouting and tube formation. Particularly in HCC, bFGF has been shown to synergistically increase the VEGF-VEGFR-2 mediated angiogenesis [38]. Moreover, a direct correlation between the levels of bFGF expression and microvessel density (MVD) in HCCs has been found [39].

The serum bFGF level correlates with advanced disease and increased rate of postoperative recurrence, however it has been demonstrated that bFGF can not be used as a single surrogate marker, since high serum levels are also found in patients with liver cirrhosis and HCC [40].

2.2.5. Thrombospondin-1

Thrombospondin-1 (TSP-1) is a large glycoprotein secreted by many cell types. It is a component of the extracellular matrix during active and subacute processes. Due to TSP-1's ability to interact with a variety of matrix proteins and cell-surface receptors, controversy exists about its conflicting functions [41,42]. In general, for the majority of the tumors TSP-1 is considered a natural angiogenesis inhibitor, inducing endothelial cell apoptosis [43]. Bocci *et al* [44] has shown that metronomic chemotherapy induces the upregulation of TSP-1 resulting in endothelial cell death. The role of TSP-1 in HCC angiogenesis is not clear. TSP-1 has been found either not to be significantly correlated with vascularity in HCC [45], or even, to be associated with invasion and tumor progression [46].

2.2.6. uPA

The urokinase plasminogen activator (uPA) system consists of the serine protease uPA, its glycolipid-anchored receptor, uPAR and its 2 serpin inhibitors, plasminogen activator inhibitor-1 (PAI-1) and plasminogen activator inhibitor-2 (PAI-2). Recent findings suggest that the uPA system is causally involved at multiple steps in cancer progression mainly in remodelling of the extracellular matrix, enhancing both endothelial and cancer cell proliferation, migration, and modulating cell adhesion [47,48]. Also, recent studies suggest that uPA expression (RNA and protein) may be a useful prognostic factor in different types of cancer [49].

The overexpression of uPA, uPAR, and PAI-1 was found in HCC, especially in the patients with portal cancer embolus, tumor invasion, and metastasis [50,51]. Immunohistochemistry results showed that the rate of positive staining of uPA, uPAR, and PAI-1 were higher in HCC than those in the control groups consisting of cancer-adjacent tissue and normal liver tissue [51]. Expression of uPA has been found to be regulated by ras and nuclear factor κB (NF-κB) signaling pathways [52]. Chan *et al* [53] has shown that uPA overexpression is associated with more aggressive HCC phenotype, and that, activation of NF-κB is closely associated with uPA overexpression.

2.2.7. Cox-2

Cyclooxygenase is a bifunctional enzyme that plays a key and rate-limiting role in the biosynthesis of prostaglandins (PGs). Cyclooxygenase catalyzes the oxygenation of arachidonic acid to the unstable product PGG_2, which is then rapidly converted to the more stable PGH_2 by the peroxidose activity of COX. PGH_2 is the common precursor of all other prostanoids, which are synthesized by distinct synthases in different tissues. COX has 2 isoforms, the constitutive isoform COX-1 and the inducible isoform COX-2. COX-1 is expressed in most normal tissues, where it mediates the PG synthesis necessary for normal physiologic functions, including platelet aggregation, glomerular blood flow regulation, and cytoprotection of the gastrointestinal mucosa [54,55]. Cox-2 expression has been identified in a number of human tumors and endothelial cells [56,57].

Several studies show that COX inhibition is capable of reducing the growth of a primary tumor, number and incidence of spontaneous metastases accompanied by increased apoptosis and decreased microvessel density [58-60]. More specifically, Cox-2 inhibition results in inhibition of synthesis of VEGF, MMP's, IL-12, factors involved in numerous angiogenic pathways [61].

Expression of COX-2 and VEGF has been observed in HCC: well-differentiated HCC expressed COX-2 more strongly than less-differentiated HCC. COX-2 expression was found to correlate with VEGF expression. Moreover, COX-2 up-regulates VEGF expression in HCC cells, possibly via PGs production [62]. Selective inhibition of COX-2 may block HCC associated angiogenesis and thus provides a rational approach for treatment of this malignancy.

Combined expression of inducible nitric oxide and COX-2 may play an important role in prognosis of HCV-positive HCC patients and that this could be partially attributable to modulation of angiogenesis by COX-2 [63].

2.2.8. Epidermal Growth Factor

EGFR belongs to the ErbB/HER protein-tyrosine kinases family, which consist of a growth-factor-binding ectodomain, a single transmembrane segment, an intracellular protein-tyrosine kinase catalytic domain, and a tyrosine-containing cytoplasmic tail [64]. The genes for the four members of this family, ErbB1–ErbB4, are found on different human chromosomes. EGFR (ErbB1) is a major transducer of mitogenic signals involved in cancer pathogenesis and progression and is considered an important target for anticancer therapy [65]. The EGFR pathway is involved in the angiogenic process by signaling through EGFR ligands, such as VEGF, TGF-a, and bFGF [66]. Conversely, selective EGFR tyrosine kinase inhibition has been shown to diminish VEGF expression and neoangiogenesis *in vitro* and *in vivo* [65-67].

More specifically, in pre clinical studies with HCC EGFR activation has been shown to result not only in cell growth but also in epidermal growth factor-induced chemotactic migration and production of active matrix MMP9 [68].

Epidermal growth factor also binds transforming growth factor alpha (TGF-alpha) which is mitogenic for hepatocytes. In a cirrosis-HCC animal model the use of Gefitinib (EFGR inhibitor) lead to decreased formation of the number of HCC nodules compare to the untreated control group, while EGFR was activated to a lesser extent in the diseased and tumoral tissues of these animals compared with untreated rats [69].

The EGFR pathway could be a potentially attractive target for this disease.

3. Angiogenesis Inhibitors-Therapeutics

3.1. Endostatin

Endostatin is the first endogenous angiogenesis inhibitor to enter human clinical trials. Endostatin is a carboxy-terminal, 20 kDa fragment of the non-collagenous (NC) 1 domain of collagen XVIII-α1. O'Reilly *et al* [70] originally reported that endostatin inhibits endothelial cell proliferation. Endostatin has subsequently been shown to affect a number of other endothelial cell functions implicated in angiogenesis. These include endothelial migration [71] ,survival , proteinase activity [72], and vessel stabilization [73].

Also, endostatin up-regulates TSP-1 while suppressing Id-1, a down-regulator of TSP-1 [74]. Other pathways found to interact to bring about a concerted antiangiogenic response following endostatin treatment include processes dictating endothelial cell fate, namely, apoptotic induction, cell cycle arrest, and decrease in motility.

Liver is the primary source for collagen XVIII, the precursor of endostatin. Recently, Graepler *et al* [75] employing an appropriate rat hepatoma model, examined the effects of sFlt-1 (soluble VEGFR-1 as an indirect inhibitor of angiogenesis) and endostatin in both single-agent as well as combined approaches under *in vitro* and *in vivo* conditions. Combined expression of sFlt-1 and endostatin effectively suppressed HCC growth under in vivo conditions. Analysis of variance did not show a significant interaction between the effects of the two factors ES(r) and sFlt-1 and their effects multiplied.

Dhar *et al* [76] has found a very strong direct correlation between VEGF and endostatin in HCC. Only the preoperative serum endostatin was found to have a significant inverse correlation with the hypervascularity. Furthermore, the combined positivity for bFGF and VEGF and negativity for endostatin was found to have a significantly positive correlation with angiogenesis. Significantly high levels of endostatin were noted in patients with trabecular-type tumors and in patients with hepatitis B infection. The serum endostatin level was found to be significantly higher in living patients and patients with high serum endostatin levels had a tendency toward long survival. This finding that was not reproduced in a more recent study by Poon *et al* [77] where endostatin was not found to be a useful prognostic marker.

Similarly in Hu *et al* [78] reported that poorly differentiated hepatoma cell lines exhibited increased endostatin/collagen XVIII levels compared with the well-differentiated ones. In hepatoma tissues, endostatin/collagen XVIII expression was detected in various types of liver cells and was significantly stronger in adjacent non-tumor tissues than that in tumors. Endostatin/collagen XVIII expression in non-tumor tissues correlated with tumor stages and expression of VEGF, but not with the stages of hepatic fibrosis.

Pan *et al* [79] investigated the inhibitory effects of endostatin-vascular endothelial growth inhibitor (VEGI151) recombinant adenoviruses on neovascularization HCC cell line and HCC-bearing nude mice model. The tumor's MVD decreased more significantly in treated mice than in AdLacZ control with an inhibition rate of 39%. Similarly recently Li *et al* [80] investigated the antitumoral activities HCC *in vitro* and *in vivo* CNHK200-mE, another E1B 55kDa-deficient adenovirus armed with a mouse endostatin gene. The recombinant adenovirus CNHK200-mE was found by this study to be a replication-competent oncolytic virus mediating high expression of therapeutic gene, capable of lysing HCC cells.

Endostatin is currently being investigated in clinical trials.

3.2. Thalidomide

The main impetus for using thalidomide to treat patients with cancer came with the discovery of its anti-angiogenic potential. The mechanism of action of thalidomide is incompletely understood. A postulated antiangiogenic effect of thalidomide was based on its ability to inhibit or interfere with integrin receptors $a_v\beta_3$ and $a_v\beta_5$ as well as inhibit bFGF and VEGF induced neovascularization [81]. Furthermore thalidomide has been found to inhibit tumor necrosis factor-α (TNF- α) acting as an antinfalammatory agent, and to provide additional indirect antiangiogenic effect since TNF- α upregulates the expression of endothelial integrin [82]. Thalidomide is administered orally daily and side effects observed are mainly fatigue, constipation and less commonly peripheral neuropathy and skin rash.

Thalidomide is one of the few agents that have been tested in the clinic as an antiangiogenic strategy against HCC. Hsu *et al* [83] reported the results of a trial with sixty-eight patients with unresectable and nonembolizable HCC. Thalidomide was given at a starting dose of 200 mg per day. The dose was gradually escalated in 100-mg steps up to 600 mg per day. One complete and 3 partial responses were seen, with a response rate of 6.3%. The duration of response was 50+, 24.6, 11.6+ and 8.7+ weeks, respectively. All 4 responders had a dramatic decrease in alpha-fetoprotein (alpha-FP) levels. Another 6 of the 42 patients with elevated alpha-FP levels before treatment had a more than 50% decrease in their alpha-FP levels after thalidomide treatment. Totally 10 patients had an objective response to thalidomide. The median overall survival for all of the 68 patients was 18.7 weeks with a 1-year survival rate of 27.6%. The median overall survival of the 10 patients with an objective response to thalidomide was 62.4 weeks. All responders responded at a dose equal to or less than 300 mg per day.

Wang *et al* [84] treated 99 patients with 150-300 mg of thalidomide. The majority had cirrotic status and comparison was made between patients who received total dose of more or less than 5000 mg total dose. Twenty-two patients were treated with dose < 5000 mg had a median survival time about 25 days, for 77 cases treated with doses equal or higher than 5000 mg the median survival time was about 109 days. Six subjects had partial response.

Similarly Lin *et al* [85] reported the results of thalidomide in 27 adult patients with unresectable HCC. Thalidomide was administered at a starting dose of 200 mg per day in a 100-mg-per-week dose escalation regimen, up to the maximum tolerated dose or to 800 mg per day. One patient experienced near-complete recovery of alpha-fetoprotein levels and a partial radiographic response on computed tomography. Two patients had stable disease during the 16-week study period. The median duration of progression-free survival was 42 days. The overall median survival was 123 days.

More recently, Patt *et al* [86] reported thirty-seven HCC patients treated with oral thalidomide, escalated from 400 mg per day during the first week to 1000 mg per day by the fifth week, delivering one-third of the dose in the morning and the remaining two-thirds of the dose in the evening prior to bedtime. One patient had a partial response (PR), 1 patient had a minor response (MR), 10 patients had stable disease (SD) (31%). The overall median survival was 6.8 months, concluding that thalidomide mostly may offer HCC patients disease stabilization.

The response rates of thalidomide trials against HCC are summarized in Table 1.

CC-5013 (Lenalidomide) belongs to a second generation thalidomide analogues with much more potent anti-inflammatory, immnumodulatory and antiangiogenic properties. Phase I clinical studies with adult patients indicate that CC-5013 does not exhibit sedation as a major side effect, but rather results in occasional neurotoxicity and myelosuppression [87]. CC-5013 is currently being evaluated in refractory solid tumors and based on its promising potent biologic activity could represent a good candidate for future trials in HCC.

Table 1. Response rate of HCC treated with antiangiogenic agents.

Treatment	N	CR/PR	OR	Ref
Thalidomide				
Patt *et al.*	37	0/2(1 MR)	5.4%	[86]
Patt *et al.*	21	0/1	4.7%	[117]
Hsu *et al.*	68	1/3	5.8%	[83]
Wang *et al.*	99	0/6	6.0%	[84]
Lin *et al.*	27	0/1	3.7%	[85]
Interferon				
Feun *et al* (+doxorubicin)	30	0/2	6.6%	[118]
Lu *et al.* (+doxorubicin+ Tamoxifen)	30	1/5	20%	[119]
Patt *et al.* (rIFNalpha2b +5-FU)	28	0/4	14.2%	[120]
Reinissh *et al.* (+rhGM-CSF)	15	0/1	6.6%	[121]
Leung *et al.* (+ Cisplatin Doxorubicin+5-FU)	148	3/22	16.8%	[101]

CR: complete response, PR: partial response, MR: minor response.

3.3. Interferon

Interferon alpha (IFN) is a cytokine with pleiotropic cellular functions, including antiviral, antiproliferative, immunomodulatory, and antiangiogenic activities [88]. Clinical studies demonstrate that IFN-a treatment can induce impressive responses in angioproliferative diseases, such as Kaposi's sarcoma [89] and hemangiomas [90]. Preclinical data suggest that these antiangiogenic activities may be associated with the regulation of endothelial cell motility [91], survival [92], and inhibition of molecules involved in the angiogenic response, such as basic fibroblast growth factor (bFGF) [93] and matrix metalloproteinase 9 (MMP-9) [94].

Interferon has been tested in HCC in both pre-clinical and clinical studies. Interferon when used in nude mice with HCC xenografts was shown, to be effective in reducing the intrahepatic recurrence rate, the size of the recurrent lesions, the rate of lung metastasis, the serum VEGF level and microvessel density, but not the bFGF levels [95]. Moreover, the combination of interferon with an angiotensin-converting enzyme inhibitor, perindopril, significantly suppressed hepatocarcinogenesis, neovascularization, and expression of VEGF. Combined treatment with IFN and perindopril resulted in a marked increase of apoptosis in the tumor, whereas tumor cell proliferation was not altered [96,97].

The activity of interferon against HCC in a clinical setting has been investigated in several studies. Lin *et al* [98] compared daily versus intermittent dosing daily 30 eligible patients were randomized into three groups: 11 patients treated with three mega units (MU) of IFN-alpha three times weekly for 24 months (IFN-alpha-continuous group), 9 patients treated with 3 MU of IFN-alpha daily for 10 days every month for 6 months followed by 3 MU of IFN-alpha daily for 10 days every 3 months for a further 18 months (IFN-alpha-intermittent group), and 10 patients who received no IFN-alpha therapy (control group). Cumulative HCC recurrence rates in the IFN-alpha-intermittent, IFN-alpha-continuous, and control groups were 22.2%, 27.3%, and 40% at the end of 1 year and 33.3%, 54.6%, and 90% at the end of 4 years.

Shiratori, *et al* [99] treated 74 patients with compensated cirrhosis, three or fewer nodules of hepatocellular carcinoma, and low hepatitis C virus RNA loads with complete ablation of lesions by percutaneous ethanol injection therapy. Forty nine patients were assigned to receive 6 million U of interferon three times weekly for 48 weeks and 25 did not receive treatment. Of the 49 patients treated with interferon, 21 showed a sustained biochemical response and 14 showed a sustained virologic response. The rate of first recurrence of new foci of hepatocellular carcinoma was similar in patients treated with interferon and untreated patients; however, the rates of second or third recurrence seemed to be lower in the interferon group than in the untreated group. Patients treated with interferon had a survival rate of 68% at 5 years and 53% at 7 years; untreated patients had a survival rate of 48% at 5 years and 23% at 7 years.

Ikeda *et al* [100] also reported cumulative recurrence rates following complete resection or ablation therapy of the treated with interferon and untreated groups: they were 0% and 62.5% at the end of the first year, and 0% and 100% at the second year, respectively in a small number of patients.

The benefit of adding intereferon to cytotoxic chemotherapy has been variable with some improvement in time to disease progression in most studies (Table 1). In one of the largest series reported by Leung *et al* [101] cisplatin, alpha-interferon, doxorubicin, and 5-fluorouracil (PIAF) was used in 148 patients. The median survival time was 30.9 weeks and significant independent predictors of an objective response were, absence of cirrhosis low bilirubin level, and positive hepatitis C serology. Good risk patients (defined by absence of cirrhosis and total bilirubin < or = 0.6mg/dL) had an objective response rate of 50%.

3.4. Bevacizumab

Bevacizumab is recombinant humanized monoclonal antibody to VEGF. Preclinical research revealed that bevacizumab specifically inhibits VEGF, has activity in multiple cancer cell lines, and is synergistic with several cancer chemotherapeutic agents [102]. Dose-limiting toxicities, the formation of antibodies to bevacizumab, and problems with wound healing after surgery have not been observed in clinical trials. A phase II study of bevacizumab in combination with 5-fluorouracil and leucovorin in patients with metastatic colorectal cancer showed promising results (i.e., therapeutic response rate and disease-free progression of survival), although a clear dose-response relationship was not demonstrated and concerns were raised about the potential for thromboembolic events, bleeding, hypertension, and proteinuria. First-line use of bevacizumab with irinotecan, 5-fluorouracil, and leucovorin produced significant improvements in response rate, duration of response, and survival in a phase III study of patients with metastatic colorectal cancer [103].

Given the preclinical data and the role of VEGF in HCC angiogenesis described above, bevacizumab is being currently tested in a Phase II randomized study in patients with unresectable HCC receiving chemoembolization and in patients with unresectable non metastatic HCC without main portal invasion.

3.5. Other Angiogenesis Inhibitors

Several other antiangiogic strategies have been investigated in animal models with HCC. Raskpof *et al* [104] employed a recombinant adenoviral vector encoding a soluble dominant negative fragment of VEGF receptor 2 (Flk-1), AdsFlk-1, to control pre-established murine orthotopic and metastatic hepatomas. Antitumoral effects of systemic AdsFlk-1 application were studied in a subcutaneous and orthotopic Hepa129 HCC model. Systemic treatment of tumor-bearing mice inhibited tumor growth by 84% compared with the corresponding control group within 16 days after vector application.

Similarly, oral administration of PTK787, a tyrosine kinase inhibitor of VEGF receptors resulted in a significant reduction in tumor volume and microvessel formation of HCC xenografts in nude mice. In addition, PTK787 induced growth arrest in HCC cells, which was associated with G1 arrest and partial G2-M block, indicating that PTK787 is a potent inhibitor of tumor growth in HCC by both antiangiogenic effect and direct effects on tumor cell proliferation and apoptosis [105].

Several new compounds targeting angiogenic pathways are under development and being investigated in clinical trials that could be used in the future against HCC. Also, known chemotherapeutics (such as the oral etoposide, oral cyclophoshamide) when are administered at relatively low, non-toxic doses on a frequent schedule, with no prolonged drug-free breaks, were found to target the tumor vasculature preferentially (metronomic chemotherapy) in several cancer preclinical models [106,107]. A possible explanation of this effect on the endothelium is the induction of TSP-1 a potent natural angiogenesis inhibitor for the majority of tumors [44]. Metronomic chemotherapy has not been evaluated yet in HCC clinical trials.

4. Challenges with Angiogenesis Inhibitors

Since the mechanism of action and the biology involved in targeting the HCC tumor vasculature is different than the traditional cytotoxic chemotherapy, the endpoints, the monitor of response and the appropriate surrogate markers are major challenges when angiogenesis inhibitors are used in the clinic. For example, classical assessment of tumor response is insufficient because a significant tumor growth delay may only occur in combination with conventional treatment. Tumor regression may also occur if there is a significant antivascular action but is often slow compared with that following conventional cytotoxic chemotherapy.

Also, measuring the tumor vascular response to antiangiogenic agents is a challenge. A direct measure of angiogenic rate would be, for example, the number of new vessels formed or the increase in vascular surface area within a tissue per day. In a non-proliferating normal tissue, this is relatively easy to measure but, in a tumor, where tumor cell proliferation accompanies angiogenesis, it is extremely difficult, even in pre-clinical models. Microvesel density (MVD) is being used commonly as an endpoint in measuring anti-angiogenesis. Unfortunately, measurement of MVD is problematic for assessing the vascular efficacy of antiangiogenic agents. As tumors grow, they can be accompanied by proportional blood vessel growth and conversely, blocking angiogenesis may be accompanied by a proportional reduction in tumor growth that would not result in a net change in MVD. Therefore, although a reduction in MVD following treatment is indicative of an antiangiogenic effect, it does not follow that no change in MVD is indicative of no antiangiogenic effect, as is commonly assumed [108].

Evaluation of human tumor angiogenesis using contrast-enhanced CT and dynamic contrast- enhanced MRI is based on the observation that angiogenesis can increase the perfusion and permeability of tumors. Dynamic contrast-enhanced magnetic resonance imaging (DCE-MRI) techniques are being used in various early clinical trials of different antiangiogenic agents. Using current imaging techniques, it is not usually possible to measure specific vascular parameters precisely [109].

Most of these techniques use analysis of the tissue uptake curves for gadolinium-conjugated DTPA (Gd-DTPA). Kanematsu M *et al* [110] investigated the correlation between magnetic resonance (MR) imaging findings and angiogenetic activity in hepatocellular nodules evaluated by immunohistochemical staining with antibody of VEGF and showed that signal intensity on unenhanced T1- and T2-weighted MR images may correlate with immunoreactivity for VEGF. Correlation was not found between immunoreactivity for VEGF and signal intensity on gadolinium-enhanced MR images. Also, Wang [111] *et al* evaluated the perfusion changes in advanced hepatocellular carcinoma (HCC) treated with thalidomide, using dynamic contrast-enhanced MRI. When comparing the MRI parameters for the tumors before, and during treatment in patients with partial response or stable disease vs the patients with progressive disease, a statistically significant difference for the peak enhancement in the first-pass study, the maximal enhancement, and the enhancement slope percentage in the first-pass study was found.

An alternative to the invasive examination of the tumor is the indirect assessment of the tumor's angiogenesis status based on the measurement of circulating biomarkers (''surrogate markers'') of angiogenesis. A number of activation-associated endothelial-derived molecules can be detected in the circulation either as specific soluble splice forms of corresponding endothelial cell surface receptors and adhesion molecules or as a consequence of proteolytic shedding of endothelial cell surface molecules [112]. As discussed previously these include the soluble forms the angiogenic endothelial cell surface receptors VEGFR-1, VEGFR-2 and Tie-2. Similarly, circulating concentrations of the angiogenic factors VEGF and bFGF are upregulated in many cancer patients. Lastly, inflammatory markers, such as sE-selectin, sICAM-1, and sVCAM-1 are presently explored as circulating biomarkers of angiogenesis [113,114]. The specificity and sensitivity of these tests has not been validated yet in large number of patients. Not only that, but other variables can affect their measurement i.e inflammation, or the fact these factors are also secreted from non endothelial cells. An emerging novel biomarker of angiogenesis is quantification of circulating endothelial progenitor cells (CEPs). CEPs are recruited from the bone marrow and can contribute to the vascularization of tumors [115]. Their precise contribution to tumor angiogenesis is presently subject of intense study. Correspondingly, the quantitative analysis of CEPs may prove to be a useful systemic biomarker of the intensity of local angiogenesis [116].

5. Expert Opinion

Several angiogenic factors have already been studied in HCC biology. Although the role of some of these factors has been well proven to either promote (i. e VEGF, bFGF, MMP's, Cox-2) or inhibit (i.e endostatin) the angiogenic process, controversy exists for others (TSP-1) despite their more clear function other tumors. Several studies have delineated possible prognostic significance of the expression of angiogenic factors and their receptorsin HCC. One of the challenges specifically related to HCC angiogenesis is the issue of associated inflammatory process encountered in patients with underlying hepatitis and/or cirrhosis, indicating that, *in vitro* angiogenic data might not be applicable to all HCC's. Even further, as the field of angiogenesis is evolving, more studies delineating possible autocrine and paracrine loops between the HCC tumor cells and the associated endothelium are required in order to translate this biology into therapeutics. For now the main antiangiogenics that have been investigated in clinical trials are interferon and thalidomide. The results of these trials have been modest, at best. The advantage of thalidomide over interferon is that is administered orally and, in general, is well tolerated, especially in patients without cirrhosis. The experience with interferon is broader than thalidomide against HCC, and when it is used with various chemotherapeutic combinations the response rate has been variable, but overall poor. Thalidomide has been studied mainly as single agent and few objective responses have been reported.

6. Five Year View

Antiangiogenic strategies against HCC are expected to expand further over the next five years. First, thalidomide and possibly its more potent analogue Lenalidomide will be further explored in combination with chemotherapeutic strategies. Second, the results of the ongoing trials with bevacizumab in HCC will determine whether the promising benefit of this agent observed with other malignancies will also be validated in HCC. Third, since in vitro studies suggest that EGFR, Cox-2, VEGFR's, MMP's are biologically attractive molecular targets for this disease trials with agents that inhibit one or more of these pathways are expected to be completed over the next few years. Fourth, inhibiting simultaneously multiple angiogenic pathways in combination with either direct angiogenesis inhibitors or even with metronomically given chemotherapy is expected to be investigated. Identifying appropriate surrogate markers of angiogenesis, and also adjusting the clinical trial design to include "biologically effective" dosing of these agents will also be challenges in the near future.

Key Issues

- Innovative approaches such as targeting the tumor vasculature are needed to treat the growing number of patients affected by HCC
- Several angiogenic pathways have been found to play crucial role in HCC biology.
- VEGF, angiopoietins, MMP's, Cox-2, bFGF, and EGF are examples of potentially attractive targets for antiangiogenic strategies against HCC.
- Interferon and Thalidomide are the antiangiogenic agents that already have been studied with variable results, without so far changing significantly the outcome of these patients.
- Bevacizumab, Lenalidomide are potent angiogenesis inhibitors being investigated currently against HCC.
- Since most of these agents inhibit single or few of the pathways involved in HCC, it remains to be determined which combination, at which disease status, with or without cytotoxic or ablative therapy, and in which order these agents have to be used in order to maximize their efficacy.
- Designing the clinical trials taking into consideration different endpoints, mechanism of action and pharmacology of these agents compare to the conventional chemotherapeutics will be one of the future challenges.

References

[1] Carr, B. I. Hepatocellular carcinoma: current management and future trends. *Gastroenterology 127*, S218-24 (2004).

[2] Cusnir, M. & Patt, Y. Z. Novel systemic therapy options for hepatocellular carcinoma. *Cancer J 10*, 97-103 (2004).

[3] Aguayo, A. & Patt, Y. Z. Nonsurgical treatment of hepatocellular carcinoma. *Semin Oncol 28*, 503-13 (2001).

[4] Tsai, H. L., Liu, C. S., Chin, T. W. & Wei, C. F. Hepatoblastoma and hepatocellular carcinoma in children. *J Chin Med Assoc 67*, 83-8 (2004).

[5] Folkman, J. Fundamental concepts of the angiogenic process. *Curr Mol Med 3*, 643-51 (2003).

[6] Folkman, J. Angiogenesis-dependent diseases. *Semin Oncol 28*, 536-42 (2001).

[7] Carmeliet, P. & Jain, R. K. Angiogenesis in cancer and other diseases. *Nature 407*, 249-57 (2000).

[8] Bergers, G. et al. Matrix metalloproteinase-9 triggers the angiogenic switch during carcinogenesis. *Nat Cell Biol 2*, 737-44 (2000).

[9] Masson, V. et al. Contribution of host MMP-2 and MMP-9 to promote tumor vascularization and invasion of malignant keratinocytes. *Faseb J 19*, 234-6 (2005).

[10] Davis, S. & Yancopoulos, G. D. The angiopoietins: Yin and Yang in angiogenesis. *Curr Top Microbiol Immunol 237*, 173-85 (1999).

[11] Heissig, B. et al. Recruitment of stem and progenitor cells from the bone marrow niche requires MMP-9 mediated release of kit-ligand. *Cell 109*, 625-37 (2002).

[12] Stupack, D. G. & Cheresh, D. A. Integrins and angiogenesis. *Curr Top Dev Biol 64*, 207-38 (2004).

[13] Bergers, G., Song, S., Meyer-Morse, N., Bergsland, E. & Hanahan, D. Benefits of targeting both pericytes and endothelial cells in the tumor vasculature with kinase inhibitors. *J Clin Invest 111*, 1287-95 (2003).

[14] Dhar, D. K. et al. Requisite role of VEGF receptors in angiogenesis of hepatocellular carcinoma: a comparison with angiopoietin/Tie pathway. *Anticancer Res 22*, 379-86 (2002).

[15] Yoshiji, H. et al. Vascular endothelial growth factor tightly regulates in vivo development of murine hepatocellular carcinoma cells. *Hepatology 28*, 1489-96 (1998).

[16] Schmitt, M., Horbach, A., Kubitz, R., Frilling, A. & Haussinger, D. Disruption of hepatocellular tight junctions by vascular endothelial growth factor (VEGF): a novel mechanism for tumor invasion. *J Hepatol 41*, 274-83 (2004).

[17] Poon, R. T., Lau, C., Yu, W. C., Fan, S. T. & Wong, J. High serum levels of vascular endothelial growth factor predict poor response to transarterial chemoembolization in hepatocellular carcinoma: a prospective study. *Oncol Rep 11*, 1077-84 (2004).

[18] Yamaguchi, R. et al. Expression and localization of vascular endothelial growth factor receptors in human hepatocellular carcinoma and non-HCC tissues. *Oncol Rep 7*, 725-9 (2000).

[19] von Marschall, Z. et al. Dual mechanism of vascular endothelial growth factor upregulation by hypoxia in human hepatocellular carcinoma. *Gut 48*, 87-96 (2001).

[20] Chow, N. H. et al. Expression of vascular endothelial growth factor in normal liver and hepatocellular carcinoma: an immunohistochemical study. *Hum Pathol 28*, 698-703 (1997).

[21] Li, X. M., Tang, Z. Y., Qin, L. X., Zhou, J. & Sun, H. C. Serum vascular endothelial growth factor is a predictor of invasion and metastasis in hepatocellular carcinoma. *J Exp Clin Cancer Res 18*, 511-7 (1999).

[22] Kim, S. J. et al. Serum vascular endothelial growth factor per platelet count in hepatocellular carcinoma: correlations with clinical parameters and survival. *Jpn J Clin Oncol 34*, 184-90 (2004).

[23] Ferrara, N. The role of VEGF in the regulation of physiological and pathological angiogenesis. *Exs*, 209-31 (2005).

[24] Holash, J. et al. Vessel cooption, regression, and growth in tumors mediated by angiopoietins and VEGF. *Science 284*, 1994-8 (1999).

[25] Suri, C. et al. Increased vascularization in mice overexpressing angiopoietin-1. *Science 282*, 468-71 (1998).

[26] Maisonpierre, P. C. et al. Angiopoietin-2, a natural antagonist for Tie2 that disrupts in vivo angiogenesis. *Science 277*, 55-60 (1997).

[27] Chen, L., Yang, Z., Wang, G. & Wang, C. Expression of angiopoietin-2 gene and its receptor Tie2 in hepatocellular carcinoma. *J Tongji Med Univ 21*, 228-30, 235 (2001).

[28] Mitsuhashi, N. et al. Angiopoietins and Tie-2 expression in angiogenesis and proliferation of human hepatocellular carcinoma. *Hepatology 37*, 1105-13 (2003).

[29] Moon, W. S. et al. Overexpression of VEGF and angiopoietin 2: a key to high vascularity of hepatocellular carcinoma? *Mod Pathol 16*, 552-7 (2003).

[30] Paradis, V. et al. A quantitative gene expression study suggests a role for angiopoietins in focal nodular hyperplasia. *Gastroenterology 124*, 651-9 (2003).

[31] Peng, L., Sun, J., Wang, W. D., Jian, Z. X. & Ou, J. R. Biological effect of ectopic expression of angiopoietin-1 and -2 in hepatocellular carcinoma cell line. *Hepatobiliary Pancreat Dis Int 2*, 94-7 (2003).

[32] Sugimachi, K. et al. Angiopoietin switching regulates angiogenesis and progression of human hepatocellular carcinoma. *J Clin Pathol 56*, 854-60 (2003).

[33] Torimura, T. et al. Overexpression of angiopoietin-1 and angiopoietin-2 in hepatocellular carcinoma. *J Hepatol 40*, 799-807 (2004).

[34] Zhao, Z. C., Zheng, S. S., Wan, Y. L., Jia, C. K. & Xie, H. Y. The molecular mechanism underlying angiogenesis in hepatocellular carcinoma: the imbalance activation of signaling pathways. *Hepatobiliary Pancreat Dis Int 2*, 529-36 (2003).

[35] Vihinen, P., Ala-Aho, R. & Kahari, V. M. Matrix metalloproteinases as therapeutic targets in cancer. *Curr Cancer Drug Targets 5*, 203-20 (2005).

[36] Visse, R. & Nagase, H. Matrix metalloproteinases and tissue inhibitors of metalloproteinases: structure, function, and biochemistry. *Circ Res 92*, 827-39 (2003).

[37] Ishii, Y., Nakasato, Y., Kobayashi, S., Yamazaki, Y. & Aoki, T. A study on angiogenesis-related matrix metalloproteinase networks in primary hepatocellular carcinoma. *J Exp Clin Cancer Res 22*, 461-70 (2003).

[38] Yoshiji, H. et al. Synergistic effect of basic fibroblast growth factor and vascular endothelial growth factor in murine hepatocellular carcinoma. *Hepatology 35*, 834-42 (2002).

[39] El-Assal, O. N., Yamanoi, A., Ono, T., Kohno, H. & Nagasue, N. The clinicopathological significance of heparanase and basic fibroblast growth factor expressions in hepatocellular carcinoma. *Clin Cancer Res 7*, 1299-305 (2001).

[40] Hsu, P. I. et al. Implications of serum basic fibroblast growth factor levels in chronic liver diseases and hepatocellular carcinoma. *Anticancer Res 17*, 2803-9 (1997).

[41] Vailhe, B. & Feige, J. J. Thrombospondins as anti-angiogenic therapeutic agents. *Curr Pharm Des 9*, 583-8 (2003).

[42] Sargiannidou, I., Qiu, C. & Tuszynski, G. P. Mechanisms of thrombospondin-1-mediated metastasis and angiogenesis. *Semin Thromb Hemost 30*, 127-36 (2004).

[43] Lawler, J. Thrombospondin-1 as an endogenous inhibitor of angiogenesis and tumor growth. *J Cell Mol Med 6*, 1-12 (2002).

[44] Bocci, G., Francia, G., Man, S., Lawler, J. & Kerbel, R. S. Thrombospondin 1, a mediator of the antiangiogenic effects of low-dose metronomic chemotherapy. *Proc Natl Acad Sci U S A 100*, 12917-22 (2003).

[45] Kawahara, N. et al. Enhanced expression of thrombospondin-1 and hypovascularity in human cholangiocarcinoma. *Hepatology 28*, 1512-7 (1998).

[46] Poon, R. T. et al. Clinical significance of thrombospondin 1 expression in hepatocellular carcinoma. *Clin Cancer Res 10*, 4150-7 (2004).

[47] Rabbani, S. A. & Mazar, A. P. The role of the plasminogen activation system in angiogenesis and metastasis. *Surg Oncol Clin N Am 10*, 393-415, x (2001).

[48] Korte, W. Changes of the coagulation and fibrinolysis system in malignancy: their possible impact on future diagnostic and therapeutic procedures. *Clin Chem Lab Med 38*, 679-92 (2000).

[49] Duffy, M. J. The urokinase plasminogen activator system: role in malignancy. *Curr Pharm Des 10*, 39-49 (2004).

[50] Qin, L. X. & Tang, Z. Y. The prognostic molecular markers in hepatocellular carcinoma. *World J Gastroenterol 8*, 385-92 (2002).

[51] Zheng, Q. et al. Invasion and metastasis of hepatocellular carcinoma in relation to urokinase-type plasminogen activator, its receptor and inhibitor. *J Cancer Res Clin Oncol 126*, 641-6 (2000).

[52] Novak, U., Cocks, B. G. & Hamilton, J. A. A labile repressor acts through the NFkB-like binding sites of the human urokinase gene. *Nucleic Acids Res 19*, 3389-93 (1991).

[53] Chan, C. F. et al. Evaluation of nuclear factor-kappaB, urokinase-type plasminogen activator, and HBx and their clinicopathological significance in hepatocellular carcinoma. *Clin Cancer Res 10*, 4140-9 (2004).

[54] Koehne, C. H. & Dubois, R. N. COX-2 inhibition and colorectal cancer. *Semin Oncol 31*, 12-21 (2004).

[55] Kujubu, D. A., Reddy, S. T., Fletcher, B. S. & Herschman, H. R. Expression of the protein product of the prostaglandin synthase-2/TIS10 gene in mitogen-stimulated Swiss 3T3 cells. *J Biol Chem 268*, 5425-30 (1993).

[56] Wang, D. & Dubois, R. N. Cyclooxygenase-2: a potential target in breast cancer. *Semin Oncol 31*, 64-73 (2004).

[57] Kim, S. K. et al. Overexpression of cyclooxygenase-2 in childhood ependymomas: role of COX-2 inhibitor in growth and multi-drug resistance in vitro. *Oncol Rep 12*, 403-9 (2004).

[58] Tortora, G. et al. Combination of a selective cyclooxygenase-2 inhibitor with epidermal growth factor receptor tyrosine kinase inhibitor ZD1839 and protein kinase A antisense causes cooperative antitumor and antiangiogenic effect. *Clin Cancer Res 9*, 1566-72 (2003).

[59] Rodrigues, S. et al. Trefoil peptides as proangiogenic factors in vivo and in vitro: implication of cyclooxygenase-2 and EGF receptor signaling. *Faseb J 17*, 7-16 (2003).

[60] Zhang, F. et al. Green fluorescent protein selectively induces HSP70-mediated up-regulation of COX-2 expression in endothelial cells. *Blood 102*, 2115-21 (2003).

[61] Gately, S. & Li, W. W. Multiple roles of COX-2 in tumor angiogenesis: a target for antiangiogenic therapy. *Semin Oncol 31*, 2-11 (2004).

[62] Cheng, A. S. et al. Cyclooxygenase-2 pathway correlates with vascular endothelial growth factor expression and tumor angiogenesis in hepatitis B virus-associated hepatocellular carcinoma. *Int J Oncol 24*, 853-60 (2004).

[63] Rahman, M. A. et al. Coexpression of inducible nitric oxide synthase and COX-2 in hepatocellular carcinoma and surrounding liver: possible involvement of COX-2 in the angiogenesis of hepatitis C virus-positive cases. *Clin Cancer Res 7*, 1325-32 (2001).

[64] Roskoski, R., Jr. The ErbB/HER receptor protein-tyrosine kinases and cancer. *Biochem Biophys Res Commun 319*, 1-11 (2004).

[65] Ciardiello, F. Epidermal growth factor receptor tyrosine kinase inhibitors as anticancer agents. *Drugs 60 Suppl 1*, 25-32; discussion 41-2 (2000).

[66] Wakeling, A. E. et al. ZD1839 (Iressa): an orally active inhibitor of epidermal growth factor signaling with potential for cancer therapy. *Cancer Res 62*, 5749-54 (2002).

[67] Ciardiello, F. & Tortora, G. A novel approach in the treatment of cancer: targeting the epidermal growth factor receptor. *Clin Cancer Res 7*, 2958-70 (2001).

[68] Matsuo, M., Sakurai, H. & Saiki, I. ZD1839, a selective epidermal growth factor receptor tyrosine kinase inhibitor, shows antimetastatic activity using a hepatocellular carcinoma model. *Mol Cancer Ther 2*, 557-61 (2003).

[69] Schiffer, E. et al. Gefitinib, an EGFR inhibitor, prevents hepatocellular carcinoma development in the rat liver with cirrhosis. *Hepatology 41*, 307-14 (2005).

[70] O'Reilly, M. S. et al. Endostatin: an endogenous inhibitor of angiogenesis and tumor growth. *Cell 88*, 277-85 (1997).

[71] Zatterstrom, U. K., Felbor, U., Fukai, N. & Olsen, B. R. Collagen XVIII/endostatin structure and functional role in angiogenesis. *Cell Struct Funct 25*, 97-101 (2000).

[72] Chavakis, E. & Dimmeler, S. Regulation of endothelial cell survival and apoptosis during angiogenesis. *Arterioscler Thromb Vasc Biol 22*, 887-93 (2002).

[73] Ergun, S. et al. Endostatin inhibits angiogenesis by stabilization of newly formed endothelial tubes. *Angiogenesis 4*, 193-206 (2001).

[74] Mazzanti, C. M. et al. Early genetic mechanisms underlying the inhibitory effects of endostatin and fumagillin on human endothelial cells. *Genome Res 14*, 1585-93 (2004).

[75] Graepler, F. et al. Combined endostatin/sFlt-1 antiangiogenic gene therapy is highly effective in a rat model of HCC. *Hepatology 41*, 879-86 (2005).

[76] Dhar, D. K. et al. Serum endostatin predicts tumor vascularity in hepatocellular carcinoma. *Cancer 95*, 2188-95 (2002).

[77] Poon, R. T. et al. Prognostic significance of serum vascular endothelial growth factor and endostatin in patients with hepatocellular carcinoma. *Br J Surg 91*, 1354-60 (2004).

[78] Hu, T. H. et al. Increased endostatin/collagen XVIII expression correlates with elevated VEGF level and poor prognosis in hepatocellular carcinoma. *Mod Pathol* (2004).

[79] Pan, X. et al. Effects of endostatin-vascular endothelial growth inhibitor chimeric recombinant adenoviruses on antiangiogenesis. *World J Gastroenterol 10*, 1409-14 (2004).

[80] Li, G. et al. Potent antitumor efficacy of an E1B 55kDa-deficient adenovirus carrying murine endostatin in hepatocellular carcinoma. *Int J Cancer 113*, 640-8 (2005).

[81] Kumar, S., Witzig, T. E. & Rajkumar, S. V. Thalidomide as an anti-cancer agent. *J Cell Mol Med 6*, 160-74 (2002).

[82] Settles, B. et al. Down-regulation of cell adhesion molecules LFA-1 and ICAM-1 after in vitro treatment with the anti-TNF-alpha agent thalidomide. *Cell Mol Biol (Noisy-le-grand) 47*, 1105-14 (2001).

[83] Hsu, C. et al. Low-dose thalidomide treatment for advanced hepatocellular carcinoma. Oncology 65, 242-9 (2003).

[84] Wang, T. E. et al. Salvage therapy for hepatocellular carcinoma with thalidomide. *World J Gastroenterol 10*, 649-53 (2004).

[85] Lin, A. Y. et al. Phase II study of thalidomide in patients with unresectable hepatocellular carcinoma. *Cancer 103*, 119-25 (2005).

[86] Patt, Y. Z. et al. Thalidomide in the treatment of patients with hepatocellular carcinoma: a phase II trial. *Cancer 103*, 749-55 (2005).

[87] Dredge, K., Dalgleish, A. G. & Marriott, J. B. Thalidomide analogs as emerging anti-cancer drugs. *Anticancer Drugs 14*, 331-5 (2003).

[88] Jonasch, E. & Haluska, F. G. Interferon in oncological practice: review of interferon biology, clinical applications, and toxicities. *Oncologist 6*, 34-55 (2001).

[89] Von Roenn, J. H. & Cianfrocca, M. Treatment of Kaposi's sarcoma. *Cancer Treat Res 104*, 127-48 (2001).

[90] Szymik-Kantorowicz, S. et al. Interferon-alpha in the treatment of high-risk haemangiomas in infants. *Eur J Pediatr Surg 15*, 11-6 (2005).

[91] Melchiori, A. et al. Interferons inhibit chemotaxis of transformed cells and their invasion of a reconstituted basement membrane. *Anticancer Res 7*, 475-9 (1987).

[92] Sgonc, R. et al. Induction of apoptosis in human dermal microvascular endothelial cells and infantile hemangiomas by interferon-alpha. *Int Arch Allergy Immunol 117*, 209-14 (1998).

[93] Dinney, C. P. et al. Inhibition of basic fibroblast growth factor expression, angiogenesis, and growth of human bladder carcinoma in mice by systemic interferon-alpha administration. *Cancer Res 58*, 808-14 (1998).

[94] Slaton, J. W., Perrotte, P., Inoue, K., Dinney, C. P. & Fidler, I. J. Interferon-alpha-mediated down-regulation of angiogenesis-related genes and therapy of bladder cancer are dependent on optimization of biological dose and schedule. *Clin Cancer Res 5*, 2726-34 (1999).

[95] Wang, L. et al. Mechanism of interferon alpha on inhibition of metastasis and angiogenesis of hepatocellular carcinoma after curative resection in nude mice. *J Gastrointest Surg 7*, 587-94 (2003).

[96] Noguchi, R. et al. Combination of interferon-beta and the angiotensin-converting enzyme inhibitor, perindopril, attenuates murine hepatocellular carcinoma development and angiogenesis. *Clin Cancer Res 9*, 6038-45 (2003).

[97] Yoshiji, H., Noguchi, R., Kuriyama, S., Yoshii, J. & Ikenaka, Y. Combination of interferon and angiotensin-converting enzyme inhibitor, perindopril, suppresses liver carcinogenesis and angiogenesis in mice. *Oncol Rep 13*, 491-5 (2005).

[98] Lin, S. M. et al. Prospective randomized controlled study of interferon-alpha in preventing hepatocellular carcinoma recurrence after medical ablation therapy for primary tumors. *Cancer 100*, 376-82 (2004).

[99] Shiratori, Y. et al. Interferon therapy after tumor ablation improves prognosis in patients with hepatocellular carcinoma associated with hepatitis C virus. *Ann Intern Med 138*, 299-306 (2003).

[100] Ikeda, K. et al. Interferon beta prevents recurrence of hepatocellular carcinoma after complete resection or ablation of the primary tumor-A prospective randomized study of hepatitis C virus-related liver cancer. *Hepatology 32*, 228-32 (2000).

[101] Leung, T. W. et al. Complete pathological remission is possible with systemic combination chemotherapy for inoperable hepatocellular carcinoma. *Clin Cancer Res 5*, 1676-81 (1999).

[102] Ignoffo, R. J. Overview of bevacizumab: a new cancer therapeutic strategy targeting vascular endothelial growth factor. *Am J Health Syst Pharm 61*, S21-6 (2004).

[103] Hurwitz, H. et al. Bevacizumab plus irinotecan, fluorouracil, and leucovorin for metastatic colorectal cancer. *N Engl J Med* 350, 2335-42 (2004).

[104] Raskopf, E. et al. Effective angiostatic treatment in a murine metastatic and orthotopic hepatoma model. *Hepatology 41*, 1233-40 (2005).

[105] Liu, Y. et al. Both antiangiogenesis- and angiogenesis-independent effects are responsible for hepatocellular carcinoma growth arrest by tyrosine kinase inhibitor PTK787/ZK222584. *Cancer Res 65*, 3691-9 (2005).

[106] Klement, G. et al. Continuous low-dose therapy with vinblastine and VEGF receptor-2 antibody induces sustained tumor regression without overt toxicity. *J Clin Invest 105*, R15-24 (2000).

[107] Browder, T. et al. Antiangiogenic scheduling of chemotherapy improves efficacy against experimental drug-resistant cancer. *Cancer Res 60*, 1878-86 (2000).

[108] Hlatky, L., Hahnfeldt, P. & Folkman, J. Clinical application of antiangiogenic therapy: microvessel density, what it does and doesn't tell us. *J Natl Cancer Inst 94*, 883-93 (2002).

[109] Miller, J. C., Pien, H. H., Sahani, D., Sorensen, A. G. & Thrall, J. H. Imaging angiogenesis: applications and potential for drug development. *J Natl Cancer Inst 97*, 172-87 (2005).

[110] Kanematsu, M. et al. Angiogenesis in hepatocellular nodules: correlation of MR imaging and vascular endothelial growth factor. *J Magn Reson Imaging 20*, 426-34 (2004).

[111] Wang, J., Chen, L. T., Tsang, Y. M., Liu, T. W. & Shih, T. T. Dynamic contrast-enhanced MRI analysis of perfusion changes in advanced hepatocellular carcinoma treated with an antiangiogenic agent: a preliminary study. *AJR Am J Roentgenol 183*, 713-9 (2004).

[112] Abdollahi, A. et al. Endostatin's antiangiogenic signaling network. *Mol Cell 13*, 649-63 (2004).

[113] Davis, D. W., McConkey, D. J., Abbruzzese, J. L. & Herbst, R. S. Surrogate markers in antiangiogenesis clinical trials. *Br J Cancer 89*, 8-14 (2003).

[114] Fox, S. B. & Harris, A. L. Markers of tumor angiogenesis: clinical applications in prognosis and anti-angiogenic therapy. *Invest New Drugs 15*, 15-28 (1997).

[115] Lyden, D. et al. Impaired recruitment of bone-marrow-derived endothelial and hematopoietic precursor cells blocks tumor angiogenesis and growth. *Nat Med 7*, 1194-201 (2001).

[116] Mancuso, P. et al. Circulating endothelial cells as a novel marker of angiogenesis. Adv *Exp Med Biol 522*, 83-97 (2003).

[117] Patt, Y. Z. et al. Durable clinical response of refractory hepatocellular carcinoma to orally administered thalidomide. *Am J Clin Oncol 23*, 319-21 (2000).

[118] Feun, L. G. et al. A phase II trial of recombinant leukocyte interferon plus doxorubicin in patients with hepatocellular carcinoma. *Am J Clin Oncol 17*, 393-5 (1994).

[119] Abramovitch, R. et al. A pivotal role of cyclic AMP-responsive element binding protein in tumor progression. *Cancer Res 64*, 1338-46 (2004).

[120] Dupuy, E. et al. Tumoral angiogenesis and tissue factor expression during hepatocellular carcinoma progression in a transgenic mouse model. *J Hepatol 38*, 793-802 (2003).

[121] Reinisch, W. et al. Prospective pilot study of recombinant granulocyte-macrophage colony-stimulating factor and interferon-gamma in patients with inoperable hepatocellular carcinoma. *J Immunother 25*, 489-99 (2002).

In: New Research on Angiogenesis Inhibitors
Editor: Elmer T. Skinard, pp. 23-54

ISBN: 1-60021-473-8

Chapter II

IS THERE A ROLE FOR ANGIOGENESIS INHIBITION IN RHEUMATOID ARTHRITIS?

***Mohammed Ali Akhavani*[1,3] *and Ewa Paleolog*[1,2*]**
[1]Kennedy Institute of Rheumatology, Faculty of Medicine, Imperial College, London, UK;
[2]Division of Surgery, Oncology, Reproductive Biology & Anaesthetics, Faculty of Medicine, Imperial College, London, UK;
[3]Restoration of Appearance and Function Trust (RAFT), The Leopold Muller Building, Mount Vernon Hospital, Northwood, Middlesex, UK.

ABSTRACT

The last decade has heralded the advent of 'biological therapies'. These treatments are specifically aimed at biologically relevant targets, rather than relieving signs and symptoms of disease, or generally (and possibly non-selectively) modifying the disease process. In general, such biological targets have been identified using a tripartite approach, involving detection of the target molecule in a given disease, identification of a clear biological and functional role for the target, and finally demonstration of the efficacy of a therapy targeting the molecule in a relevant disease system. The fundamental model of a biological therapy is inhibition of tumour necrosis factor (TNF) α. The seminal studies demonstrating that TNFα is expressed in rheumatoid arthritis (RA) and that blockade of TNFα reduced disease severity in an *in vivo* model eventually resulted in the approval of anti-TNFα for the treatment of RA.

In oncology, biological therapies frequently target the vasculature, which fulfils a gatekeeper role in cell recruitment, through expression of leukocyte activators and chemoattractants, as well as adhesion molecules. Endothelial cells also play a role in the

* Correspondence concerning this article should be addressed to Dr. Ewa Paleolog, Kennedy Institute of Rheumatology, Faculty of Medicine, Imperial College, Arthritis Research Campaign Building, 1 Aspenlea Road, London W6 8LH, United Kingdom. Tel: 44-20-8383-4481; Fax: 44-20-8383-4499; E-mail: e.paleolog@imperial.ac.uk.

growth of new blood vessels, termed 'angiogenesis', which is vital for efficient supply of oxygen and nutrients to tissue. Angiogenesis has been a putative target in cancer since it was first linked to tumour growth and metastases in the 1970s. The most successful anti-angiogenic approach is antibody targeting the angiogenic stimulus vascular endothelial growth factor (VEGF). This antibody, bevacizumab, was approved in combination with chemotherapy for patients with colorectal cancer.

However, biological therapies targeting angiogenesis may have a broader application in the clinic. In RA, the expansion of the synovial tissue necessitates an alteration in the synovial vascular supply, to cope with the increased requirement for oxygen and nutrients. VEGF and many other pro-angiogenic factors have been demonstrated to be expressed in RA. Several studies have shown that targeting angiogenesis in animal models of arthritis ameliorates disease. It is therefore likely that, as in the oncological setting, angiogenesis is fundamental in RA and represents a potentially attractive target for therapy.

While the impact of TNFα inhibition on management of RA cannot be understated, the incidence of infections, and the lack of responsiveness in certain patients, supports the need for further therapeutic options. The focus of this chapter is to discuss the role of the vasculature in RA, particularly in the light of observations using therapies such as anti-TNFα, and on the potential for development of vascular-targeted therapies for treatment of RA. Furthermore, the potential of approaches targeting VEGF, or other angiogenic factors, will be discussed.

ABBREVIATIONS

CIA	collagen-induced arthritis
DMARD	disease-modifying anti-rheumatic drugs
EGF	epidermal growth factor
FGF	fibroblast growth factor
HIF	hypoxia-inducible factor
HRE	hypoxia-responsive element
IL	interleukin
MMP	metalloproteinase
MTX	methotrexate
PDGF	platelet-derived growth factor
PlGF	placental growth factor
RA	rheumatoid arthritis
TGF	transforming growth factor
TIMP	tissue inhibitor of metalloproteinases
TNFα	tumour necrosis factor α
VEGF	vascular endothelial growth factor
VEGF-R	VEGF receptor

INTRODUCTION

Rheumatoid arthritis (RA) is a chronic autoimmune inflammatory disease of unknown aetiology, with a prevalence of about 1% in most parts of the world. It is generally considered that interactions between genetic factors, sex hormones, and possibly an infectious agent or another immune activating factor, initiates an autoimmune mechanism that culminates in a disease with inflammatory and destructive features. RA is characterized by deforming symmetrical polyarthritis, with loss of articular cartilage and erosion of juxta-articular bone. In a proportion of patients systemic and extra-articular features may be observed, including anaemia, vasculitis, nodules in subcutaneous, pulmonary and sclera tissues, and interstitial inflammation in lungs as well as in exocrine salivary and lachrymal tissue. Approximately half of RA patients have tendon involvement, and dorsal wrist swelling due to tenosynovitis is often the first presentation of the disease [1]. Proliferation of the synovial lining of tendons causes scarring and adhesion formation and 50% of patients with tendon disease will also show tenosynovial invasion into the tendon substance itself. This invasion is associated with multiple tendon ruptures and a poorer prognosis for long-term hand function [2]. Furthermore, RA is associated with increased mortality, most probably due to the high frequency of cardiovascular disease [3-5]. A recent study from Finland also reported that RA patients were at increased risk of dying of malignancies, as well as urogenital, gastrointestinal and respiratory diseases [6].

At the cellular level, RA is characterised by chronic inflammation of synovial tissue lining the joints and tendons, with synovial proliferation and infiltration by blood derived cells. Normally the synovium is made up of a well-organized matrix containing proteoglycan aggregates. Within this structure are found the synovial cells, as well as a network of capillaries and lymphatic vessels. Between the cartilage and synovium is the synovial fluid, which nourishes and lubricates the joint. However, in RA the synovium becomes infiltrated by cells of lympho-haematopoietic origin, chiefly T cells, B cells and macrophages. The synovial fluid becomes rich in neutrophils, and increases in volume due to oedema, leading to joint swelling and pain. In addition, the synovium becomes thickened, from a layer of 1-2 cells to approximately 6-8 cells, and becomes locally invasive at the interface with cartilage and bone or tendon. Progressive destruction of underlying tissue eventually leads to produce deformities and functional deterioration in the long term [7].

In the past, the traditional treatment of RA was represented by a pyramidal approach starting with non-steroidal anti-inflammatory drugs at the base of the pyramid and progressing to disease modifying anti-rheumatic drugs (DMARD) such as gold, sulphasalazine and methotrexate (MTX). Despite such pharmacological interventions, the majority of patients with a more aggressive disease evolution become clinically disabled within 20 years. Furthermore, up to 90% of patients with aggressive synovitis have radiological evidence of bone erosion within 2 years of diagnosis, despite treatment. However, over the last 15 years, advances in the understanding of the pathogenesis of RA, based on studies of human tissues and animal models of disease, have led to the identification of a number of new molecular targets for intervention. The first of these was tumour necrosis factor α (TNFα), which mediates many inflammatory and immunoregulatory activities relevant to the development of RA, for example upregulation of adhesion molecules for leukocytes on the endothelial lining of blood vessels, and induction of pro-inflammatory cytokines, such as interlukin (IL)-1 and IL-6. Additional activities include induction of chemokines and metalloproteinase (MMP) enzymes that mediate bone and cartilage destruction. TNFα has been detected in synovial fluid, and is produced in by enzymatically dissociated synovial cells from RA patients [8]. Immunohistochemical analyses have demonstrated the presence of TNFα, together with that of its receptors in RA synovium. Inhibition of TNFα activity in synovial joint cell cultures resulted in the downregulation of release of other cytokines, such as IL-1, IL-6 and IL-8 [8]. Administration of anti-TNFα antibody in the murine model of collagen-induced arthritis (CIA), which has many similarities to human RA, reduced the severity of disease [9]. The observation of TNFα expression in RA, inhibition by TNFα antibody of cytokine production by synovial cells, and effectiveness of TNFα blockade in murine arthritis, formed the basis of the hypothesis that TNFα was a possible therapeutic target in RA [10].

To date three biological inhibitors of this cytokine have been approved for clinical use, and more than 1,000,000 patients have been treated with these inhibitors. Monoclonal antibodies such as infliximab and adalimumab bind to TNFα with a high affinity. In contrast, etanercept is a fusion protein containing soluble TNF receptor type II fragments linked to an immunoglobulin fragment. Infliximab was the first biological developed, and from the earliest trials has shown remarkable therapeutic efficacy [11,12]. However, whereas TNFα inhibitors such as infliximab, etanercept and adalimumab have demonstrated an acceptable safety profile, there is nonetheless an increased risk of minor infection and, more rarely, of serious infection such as tuberculosis [13]. Furthermore, targeting combinations of cytokines, for example blockade of IL-1 plus TNFα, while effective in animal models of arthritis, is likely to be pro-infective in RA. This was the case in a small combination trial of etanercept and anakinra [14]. Thus, despite the clinical success of anti-cytokine biologicals, further initiatives in drug discovery are highly desirable with a view to further improvements in the pharmacological management of RA.

Angiogenesis: a CENTRAL Role in RA

Formation of new blood vessels from pre-existing vasculature ('angiogenesis') is a fundamental biological process, important from the earliest stages of embryo development, right through to the adult, playing a key role in physiological processes such as healing of wounds or fractures and in the female reproductive cycle [15]. Organization of blood vessels, in terms of time, space and composition, needs to take into account the requirements of a particular organ or tissue, and appropriately adapt the type of mural cells and composition of sub-endothelial matrix. The process of angiogenesis generally commences with degradation of the basement membrane of existing vessels by proteolytic enzymes such as MMP, to allow endothelial cells to proliferate and to migrate in a directional manner towards an angiogenic stimulus. A provisional extracellular matrix is laid down, consisting of molecules such as fibrin and fibronectin. Cell adhesion molecules mediate the interaction of the endothelial cells in the newly formed sprout, as they adhere to one another and to the extracellular matrix. Subsequently these endothelial sprouts differentiate into mature vessels, which become stabilised and surrounded by pericytes. Finally the processes of lumen formation, capillary loop formation and stabilisation of the mature vessels occur. The endothelial cells orientate themselves so that the luminal surfaces are aligned creating vessels, which then branch to form the meshwork of capillary loops. Other stimulatory molecules recruit mesenchymal cells which differentiate into smooth muscle like-pericytes and surround the mature vasculature [16]. This cascade of tightly regulated events leads to construction of a new vascular tree, evidenced in the physiological setting in embryonic development, reproduction and healing. Unregulated angiogenesis however, leads to pathological conditions such as RA and cancer.

A range of different factors can promote angiogenesis, directly or indirectly, including fibroblast growth factor (FGF)-1 and FGF-2, angiopoietins, epidermal growth factor (EGF), platelet-derived growth factor (PDGF) and transforming growth factor (TGF) β. Various angiogenesis inhibitors have also been described, with angiostatin and endostatin being the most notable. However, the best characterised pro-angiogenic stimulus vascular endothelial growth factor (VEGF) [17-19]. The VEGF family now contains at least 6 related cytokines, although the original member, VEGF, remains the most extensively studied. VEGF is secreted by various cell types *in vitro*, including fibroblasts, macrophages, platelets and lymphocytes, and expression is elevated in a range of angiogenesis-associated disease states. The gene for human VEGF is located on chromosome 6p21.3 and organized into 8 exons, separated by 7 introns. The resultant mRNA undergoes alternative splicing events to generate several transcripts. The most prevalent form, VEGF-165, contains 165 amino acids encoded for by exons 1-5 and 7-8, but lacks the region encoded by exon 6. VEGF-121 is encoded for by exons 1-5 and 8, with VEGF-189 and VEGF-206 also including regions encoded for by exon 6. These various VEGF isoforms exhibit different heparin-binding properties, which govern whether the different glycoproteins are secreted or remain cell-associated. VEGF-165 and VEGF-121 are secreted efficiently from producing cells. This contrasts with VEGF-189 and VEGF-206, which remain bound to the cell surface and extracellular matrix, presumably due to the presence of a highly cationic 24 amino acid sequence encoded by exon 6. Two VEGF receptors (VEGF-R) have been identified, VEGF-R1 (Flt-1) and VEGF-R2 (KDR/Flk-1) [17]. VEGF is over-expressed by numerous solid angiogenic tumours and haematological malignancies, and interrupting the VEGF pathway has become a major focus of oncological research [20]. The most successful anti-angiogenic approach is bevacizumab, a humanised monoclonal anti-VEGF antibody. A phase III study in metastatic colorectal cancer demonstrated increased response rates, prolonged time to progression and significantly increased survival times when bevacizumab was used with first-line chemotherapy [21], and bevacizumab is now approved in combination with 5-Fluorouracil-based chemotherapy as first- or second-line treatment of patients with metastatic carcinoma of the colon or rectum. Other approaches to block angiogenesis are based on VEGF receptors, such as high affinity 'decoy' receptors known as 'VEGF Traps' or anti-VEGF receptor antibodies. Other approaches have targeted the tyrosine kinase activity of VEGF receptors.

Angiogenesis is thus an emerging target for therapy. However while much of the focus of angiogenesis-related research has been on modulating angiogenesis in cancer, RA is also a possible target for angiogenesis blockade. The principle for angiogenesis blockade for treatment of cancer has always been to 'starve' the tumour of nutrients and oxygen. A feature of RA is the hyperplastic, inflammatory and erosive synovium, be it in the context of joint or soft tissue disease. This attribute implies that a number of concurrent processes must be occurring, namely cellular proliferation, changes in cell metabolism, inflammatory cell trafficking, alterations in matrix components and generation of signals promoting cell motility. Blood vessels fulfil an integral role in all of these processes, by supplying oxygen and nutrients necessary for cell metabolism and division, as well as by bringing in leukocytes and signalling mediators such as cytokines, chemoattractants and growth factors. As the synovium expands, more blood vessels are needed, to supply poorly perfused and oxygenated areas distant from the pre-existing blood vessels. Thus the RA synovium exhibits tumour-like features of hypoxia, angiogenesis, inflammation and invasiveness.

One of the earliest reports suggesting that RA is coupled with a pro-angiogenic switch was the description in 1980 that synovial fluids from patients with RA contained a low molecular weight angiogenesis factor [22]. RA synovial fluids were later shown to induce morphological changes in human endothelial cells, with formation of tubule-like structures and induction of angiogenesis in an *in vitro* assay [23,24]. It is now clear that RA synovium is characterised by an abundance of blood vessels, of different sizes, both in areas of diffuse synovitis and in regions of large leukocytic infiltrates with germinal centre-like structures [25]. Alterations in endothelial cell turnover and apoptosis have been reported, with expression cycle-associated antigens, including Ki67, and integrin $\alpha V\beta 3$, which is associated with vascular proliferation [26-28].

Expression of protein and/or mRNA for many angiogenic growth factors has been described in RA synovium. More than 10 years ago the groups of Koch and Fava almost simultaneously reported VEGF expression in RA synovial fluids and tissue [29,30]. Since these original studies, there have been many other reports documenting expression of VEGF and its receptors not only in RA, but also in other arthritides. VEGF expression is also increased in juvenile RA [31,32], systemic sclerosis [33], ankylosing spondylitis [34,35] and psoriatic arthritis [25,28,36]. In RA, circulating (serum) levels of VEGF are increased, and correlate with inflammatory response markers such as C-reactive protein and swollen joint counts [28,33,37-40]. VEGF levels are increased even in RA patients with disease duration of less than 2 years, and predict subsequent joint destruction, suggesting that angiogenesis may be an early event in RA progression [36,41]. Indeed, we have reported a highly significant correlation between serum VEGF levels at presentation in untreated RA patients, and the change in hand and feet radiographs, taken at initial presentation and at 1 year follow-up and scored according to the van der Heijde modification of Sharp's method [36]. Treatment of RA with TNFα inhibitors (alone or with MTX), DMARD or anti-IL-6 receptor antibody significantly reduced serum VEGF concentrations [36,38,42-46]. For example, a marked reduction in serum VEGF levels was seen even 4 weeks after a single infusion of anti-TNFα antibody infliximab, and this reduction correlated with alterations in disease parameters such as C-reactive protein levels and swollen joint counts. Treatment of RA patients with a combination of multiple infusions of infliximab and MTX resulted in a more prolonged decrease in serum VEGF levels relative to patients who received infliximab without MTX [38]. Serum VEGF levels are also apparently higher in patients with follicular synovitis (with germinal centre-like structures), as opposed to diffuse synovitis [47], and in patients with extra-articular manifestations of RA [48].

VEGF expression is localised to synovial macrophages, neutrophils and fibroblasts [49-51]. Stimuli for VEGF release include pro-inflammatory cytokines such as IL-1 [38,52,53], TNFα [38,54] and TGFβ [55], as well as hypoxia, which will be discussed in more detail later. VEGF is not only expressed in RA joint synovium, but also in synovium invading tendons, suggesting that angiogenesis may have a role in tenosynovial proliferation and tendon destruction [56]. VEGF isoforms VEGF-165 and VEGF-121 appear to be the predominant forms expressed [57]. The biological significance of VEGF expression in RA synovium was confirmed by the observation that VEGF-R1 and VEGF-R2 mRNA was expressed by RA synovial microvascular endothelial cells of nearby blood vessels [30], and that conditioned medium from synovial tissue explants was mitogenic for endothelial cells, an activity reduced by anti-VEGF antibody [29]. Another study utilised an antibody which recognises VEGF-R2 when complexed with VEGF, and found that expression was higher in RA synovium relative to control tissue [58]. A sub-set of VEGF binding molecules are the semaphorin receptors neuropilin-1 and neuropilin-2. Neuropilin-1 has been shown to bind VEGF-165 and thereby enhance VEGF-R2-mediated signal transduction. Neuropilin-1 is also expressed in RA, suggesting that neuropilin-1 may amplify VEGF signalling through VEGF-R2 in RA [59].

At this point it is worth noting that while VEGF is traditionally considered for its role in angiogenesis, there may be another important function for this growth factor in terms of blood vessel development. In the embryo, development of the vascular system occurs via a combination of 2 distinct processes, vasculogenesis (coalescence of endothelial cells) and angiogenesis. Formation of the primordial vascular network results from the commitment by precursor cells (termed haemangioblasts) in the embryonic and extra-embryonic mesoderm to form endothelial cells, rather than haematopoietic cells. Post-natally, vasculogenesis has now been shown to be recapitulated in adults, especially in the pathological setting. Endothelial progenitor cells isolated from human peripheral blood have been found to differentiate into endothelial cells and express both haematopoietic markers such as CD133, and endothelial markers, including CD34, CD31 and VEGF-R2. Crucially, these cells also incorporate into sites of angiogenesis *in vivo* [60]. This was further expanded upon in studies which demonstrated that VEGF can mobilise endothelial progenitors from the bone marrow [61]. In RA, CD133/CD34/VEGF-R2-positive cells were found in apposition to RA synovial blood vessels [62]. Interestingly, these cells were generated at a higher rate from bone marrow samples taken from RA patients, compared to normal subjects. Furthermore, the capacity of bone marrow-derived cells from RA patients to progress into endothelial cells correlated with the synovial microvessel density [63]. Circulating endothelial progenitor cells in patients with active RA have been reported to be lower than in individuals with inactive disease or in healthy controls [64]. It therefore seems likely that increased synovial vessel density in RA is not just due to angiogenesis, but also results from post-natal vasculogenesis, due to increased mobilisation of endothelial progenitors from the bone marrow, potentially driven by VEGF and/or hypoxia, and increased endothelial progenitor cell homing to the synovium.

In addition to VEGF, numerous other angiogenic factors are expressed in RA synovium. For example, expression of angiopoietin-1 and angiopoietin-2 in RA synovial tissue has been described [65,66], together with their receptors Tie-1 and Tie-2 [67,68]. An interesting study documented that expression of angiopoietin-2 and VEGF was higher in synovium of patients with psoriatic arthritis, relative to RA, whereas angiopoietin-1 levels were more comparable. Psoriatic arthritis and RA exhibit different features in terms of vascular morphology, in that blood vessels in psoriatic synovium were highly tortuous in appearance, compared to the predominantly straight and branching vessels seen in RA, suggesting that the balance between angiopoietin-1, angiopoietin-2 and VEGF may affect vessel growth and maturation in arthritic synovium [25]. It has been suggested that synovial VEGF expression is highest in patients whose vascular morphology is classified as straight, rather than tortuous or a mixture of both [69]. FGF-1 and FGF-2 have been detected in RA synovial tissue [70-72], as well as PDGF [71,73] and hepatocyte growth factor [74]. The potential role of TGFβ in synovial angiogenesis is unclear. TGFβ was shown to induce VEGF expression in human synovial fibroblasts [75]. Indeed, TGFβ is by far the most powerful inducer of VEGF secretion by human synovial fibroblasts, when compared with other cytokines associated with the pathogenesis of RA, such as IL-1 or PDGF. Thus, it seems likely that in RA, TGFβ exerts its angiogenic effects predominantly through the induction of VEGF secretion by fibroblasts.

Regulation of Angiogenesis by Hypoxia

Although induction of VEGF expression by cytokines such as TNFα is likely to be important in RA synovial angiogenesis, the foremost driving force for angiogenesis in the context of RA appears to be the need to normalise perfusion and oxygenation of the ever-expanding and highly metabolically active synovial mass. As in the case of tumour growth, the mass of the hyperplastic RA pannus may result in local hypoxia, thus driving the need for a compensatory neovascularisation, to increase the supply of nutrients and oxygen [76]. Decreased synovial oxygen tension has been reported in RA [77]. Similarly in an animal model of arthritis, onset of disease was associated with a marked reduction in synovial oxygen tensions [78]. Several studies have demonstrated that the oxygen consumption of the RA synovium is elevated, possibly due to the increased proliferative activity of synovial cells, and that glucose is oxidized via an anaerobic, rather than aerobic, pathway [79-81]. Upregulation in RA synovium of hypoxia-sensitive glycolytic enzymes such as glucose-6-phosphate isomerase has been described [82], and this phenomenon of hypoxia within the RA joint is supported by observations of decreased synovial fluid oxygen tensions [76]. Moreover, the inflamed joints of arthritic individuals are unavoidably subjected to movement. Accumulation of fluid within the joints may also increase intra-articular pressure and temporarily obliterate capillary flow in the synovium, compounding the reduced perfusion still further.

The ability of cells to adapt to periods of hypoxia is important for their survival. A 'master regulator' of the adaptive response to alterations in oxygen tension is hypoxia-inducible factor (HIF), a transcriptional complex containing two (α and β) members of the basic-Helix-Loop-Helix Per-Arnt-Sim family [83]. Under normoxic conditions, HIF-α sub-units have a short half-life, due to hydroxylation of proline residues 402 and 564 by prolyl 4-hydroxylase enzymes, which require oxygen as a co-substrate [84]. This allows binding of von Hippel-Lindau tumour suppressor protein, which targets HIF-α for polyubiquitination by ubiquitin E3 ligases. Once polyubiquitinated, HIF-α-subunits are degraded by the 26Sproteasome [85]. Additional oxygen-dependent hydroxylation of asparagine 803 within HIF-α regulates recruitment of transcriptional co-activators such as p300/CBP [86].

The dependence of prolyl and asparaginyl hydroxylation on oxygen means that under conditions of hypoxia, HIF-α accumulates in the nucleus where, upon binding to HIF-1β and recruitment of p300/CBP, it recognizes specific hypoxia-responsive elements (HRE) within promoters of target genes, leading to their transactivation [87,88]. HRE have been found in more than 60 genes, which include erythropoietin, glycolytic enzymes, glucose transporters such as GLUT-1 and genes involved in iron metabolism and cell survival [89]. Two transactivating domains have been identified in the C-terminal part of the HIF-1α protein, namely an N-terminal transactivation domain, which overlaps with oxygen dependent degradation domain, and the more downstream C-terminal transactivation domain, which interacts with co-activators such as p300/CBP to activate transcription. Further upstream of the transactivating domains, a contiguous basic-Helix-Loop-Helix Per-Arnt-Sim domain creates a functional interface for subunit dimerisation and binding to HRE [88].

Several studies have shown that hypoxia is a potent stimulus for VEGF induction in RA synovial membrane cells [38,90] and in RA fibroblasts [53,55]. HIF-1α, HIF-2α and VEGF are expressed in RA synovial macrophages and fibroblasts [91-93], as well as in experimental arthritis models [94], suggesting that synovial hypoxia leads to upregulation of HIF in the joint, accumulation of VEGF and induction of synovial angiogenesis. We have observed that HIF-1α and HIF-2α are not only expressed in human RA joint synovium, but also in RA tenosynovium. HIF-2α appears to be consistently expressed in the synovial intimal lining layer, as well as within areas of cellular infiltrates deeper within the synovium. In contrast, HIF-1α is expressed in particular by the endothelial cells lining blood vessels in areas of diffuse synovitis. VEGF expression appears to more closely resemble that of HIF-1α, rather than HIF-2α [95]. Targeted deletion of HIF-1α in myeloid cells resulted in the inability of macrophages to upregulate molecules involved in glycolysis (such as GLUT-1) and impaired macrophage aggregation and migration. In a model of arthritis in which disease is induced by serum from a strain-specific arthritis model, loss of myeloid cell HIF-1α abolished paw swelling and joint destruction [96].

Finally, as mentioned earlier, blood vessel formation may also arise from recruitment of bone marrow-derived endothelial progenitors, and induction of other HIF-1 targets may be relevant in this situation. Stromal cell-derived factor 1 (SDF-1 or CXCL-12) is a prime example. This chemokine is regulated by HIF-1 [91,97], and has been hypothesised to mobilise endothelial progenitor cells from the bone marrow [98]. Coupled with the observation of increased generation endothelial progenitor cells from bone marrow samples taken from RA patients, compared to normal subjects [63], these data suggest that RA tissue could represent a conditional stem cell niche, with recruitment and retention of circulating endothelial progenitor cells via a hypoxia-sensitive mechanisms, and concomitant vasculogenesis and angiogenesis.

Hypoxia thus appears to be an important regulator of synovial angiogenesis, principally through induction of VEGF. Almost certainly alternative hypoxia-regulated mechanisms will also contribute to the invasive capacity of endothelial and/or synovial cells, and these will be discussed in more detail in the next section.

MMP and RA: Dual Role in Angiogenesis and in Synovial Invasion

In RA, joint destruction occurs when proliferating mass of RA synovial tissue ('pannus'), is in contact with the intra-articular cartilage and the bone surface [99]. This synovial mass has tendency to invade the superficial layers of the joint cartilage. Cells such as macrophages, neutrophils and fibroblasts are found at the pannus/cartilage border. However the precise mechanism of how these cells contribute to the observed cartilage breakdown is uncertain. Furthermore, while it is known that a combination of mechanical attrition and ischemia, together with synovial proliferation and invasion, causes tendons to rupture in RA, the exact mechanism by which the synovium brings about this destruction is unknown. It is important to note that, histologically, invasive tenosynovium has the characteristics of chronic inflammatory tissue, including RA synovium invading cartilage and bone. At the electron microscopic level, the affected tendons show an alteration in collagen structure, with an increased number of thin collagen fibrils and degenerative nodules known as rice bodies. While there are reports regarding the histological changes to the tendon in RA, surprisingly little is known regarding the involvement of pro-inflammatory cytokines or proteolytic enzymes in this process. It has, however, been demonstrated that tendon tissue cells can actively secrete MMP and tissue inhibitors of MMP (TIMP), suggesting that tendon is capable of producing a remodelling response [100]. MMP are a family of over 20 zinc-containing endopeptidases that are capable of degrading various components of the extracellular matrix [101,102]. All are produced as latent pro-enzymes which must be proteolytically processed to be activated. MMP can be divided into at least five sub-groups, based on their structure and/or substrate specificities, namely the matrilysins, collagenases, gelatinases, stromelysins and membrane-type MMP (MT-MMP) [103]. This section will discuss the function of MMP in RA, focusing on the involvement of hypoxia in the dual regulation of angiogenesis and destruction of cartilage and other components of connective tissue.

It is considered that an imbalance between excessive levels of many of the MMP, coupled with inadequate TIMP levels, facilitates the joint destructive process, and this has been demonstrated in RA cartilage extracts [104]. The collagenase sub-group of MMP consists of MMP-1 (collagenase 1), MMP-8 (collagenase 2), and MMP-13 (collagenase 3), all of which are capable of degrading collagen types I-IV (including collagen II, which is found in cartilage) and type VII, the primary component of tendon. MMP-1, MMP-8 and MMP-13 are present in synovial fluids and synovial membranes of RA patients [105-107]. MMP-8 has been reported to be expressed by RA synovial fibroblasts and endothelial cells [108]. MMP-3 (stromelysin 1) is unable to degrade type I collagen, but is important in cartilage destruction [109]. Gelatinases (MMP-2 and MMP-9) have an additional fibronectin domain located inside the catalytic domain, and efficiently digest denatured collagen, or gelatin. MMP-2 (gelatinase A) degrades types IV, V, VII, and X collagen, elastin, fibronectin, laminin and several chemokines, and activates proMMP-1 and proMMP-9. It is constitutively expressed by many cells, and has important functions in cell migration [110]. MMP-9, or gelatinase B, efficiently degrades denatured collagen, types IV, V, and XI collagen, the N-telopeptides of type I collagen, aggrecan, the cartilage link protein, and elastin [111]. MMP-9 is also present in the synovial fluid, synovial membrane, and serum of

RA patients [106,112], suggesting its importance in this disease. Elevated activation of MMP-2 and MMP-9 in RA fibroblasts following co-culture with monocytic cells has been described. This has been suggested to involve CD147, also known as extracellular MMP inducer or EMMPRIN [113]. MMP are also implicated in tendon destruction. Rheumatoid tenosynovium produces biologically active MMP enzymes (MMP-1, -2, -8 and -13), capable of cleaving type I collagen of tendon, and indeed invasive tenosynovium produces higher amounts of these enzymes compared to non-invasive tenosynovium [56,114]. Jain *et al* also showed similar levels of TIMP-1 in the different tenosynovial sample groups, suggesting that the increase in MMP production in the invasive *versus* the non-invasive tenosynovial tissue is not compensated sufficiently by an increase in TIMP-1 production [56].

The MT-MMP are expressed on the cell surface as either a type I transmembrane protein [115] or a glycophosphatidylinositol-anchored protein [116]. As a result, they can act only in the pericellular space (unlike soluble MMP which are secreted from cells into the extracellular milieu and can diffuse to distal sites), thus spatially restricting matrix degradation to the immediate vicinity of the cell. MT1-MMP is one of the better studied of the MT-MMP, and has an important function in degrading collagen. Collagen is resistant to most proteinases at neutral pH apart from collagenolytic MMP (i.e. MMP-1, MMP-2, MMP-8, MMP-13, and MT1-MMP) [111]. As well as collagen (types I, II and III), MT1-MMP also cleaves other extracellular matrix substrates including fibronectins and laminins [117]. Additionally, MT1-MMP is required to activate proMMP-2, which leads to production of MMP-2 that can also degrade type IV collagen, thus allowing cells to traverse the basement membrane [118]. The importance of MT1-MMP in bone and cartilage turnover is highlighted by MT1-MMP knock out mice [119]. These mice initially show no abnormal phenotype but fail to have an appropriate growth pattern, develop severe fibrosis in periskeletal soft tissue and show delayed ossification of bone. Eventually they die after a period of about ten weeks [120].

As well as regulation at the level of transcription, MMP activity is also regulated by endogenous inhibitors, primarily TIMP, which have been identified in species ranging from drosophila to zebrafish, suggesting that these proteins are ancient eukaryotic proteins [121-123]. Four TIMP have been characterized to date in humans, designated TIMP-1, -2, -3 and -4 and expressed by a variety of cell types and present in most tissues and body fluids. TIMP are natural MMP inhibitors and inhibit the MMP proteolytic activity by forming non-covalent 1:1 stoichiometric complexes that are resistant to heat denaturation and proteolytic degradation [124]. Although they all inhibit MMP proteolytic activity, TIMP differ in many aspects including solubility, interaction with proenzymes and regulation of their expression. TIMP-1, TIMP-2 and TIMP-4 are present in soluble forms while TIMP-3 is tightly bound to the matrix. TIMP bind to MMP but they have also the ability to form complexes with proMMP, thus regulating the MMP activation process. TIMP-1 forms a preferential complex with proMMP-9 [125]. While TIMP-2 binds to proMMP-2 and facilitates enzyme activation [126], TIMP-3 binds with both proMMP-2 and proMMP-9 [127]. TIMP-4 can also bind to the C-terminal domain of proMMP-2 [128].

Their function in breakdown of extracellular matrix means that MMP are likely to be critical for angiogenesis [129], but because of overlapping activities, most MMP knockout mice have no obvious angiogenic phenotype. The exception to this is the MT1-MMP knockout mouse, which shows severe defects in angiogenesis, as well as craniofacial, axial and appendicular skeletal defects. These defects were in part due to abnormal bone growth, as a result of defective vascular invasion of cartilage, enlargement of the hypertrophic area of the growth plate and decreased chondrocyte proliferation [130]. Expression of MT1-MMP in tumours has been shown to fuel angiogenesis *in vivo* by stimulating VEGF-A synthesis from tumour cells [131,132]. Transfection of endothelial cells with a vector expressing MT1-MMP was shown to augment MMP-2 activation, together with migration of endothelial cells and formation of tubule-like structures [133]. Subsequently, MT1-MMP was shown to be involved in endothelial migration and invasion of fibrin and collagen [134-138].

Other MMP have also been shown to participate in angiogenesis. Invasion of collagen gels and formation of capillary networks is associated with MMP-1 expression [139]. MMP-7 (matrilysin) boosts endothelial cell proliferation, up-regulates endothelial expression of MMP-1 and MMP-2, and induces angiogenesis *in vivo* [140,141]. Proteolytic breakdown of type IV collagen by MMP opens the cryptic $\alpha v\beta 3$ integrin binding sites, which encourages angiogenesis. This cleavage is associated with a loss of binding to integrin $\alpha 1\beta 1$ and an enhancement in $\alpha v\beta 3$ binding, and correlates with increased MMP-2 expression and activation. Exposure of these cryptic sites is found within the endothelial basement membranes in angiogenic and tumour blood vessels, but not in inactive vessels [142]. MMP-2 binding *via* its PEX domain to $\alpha v\beta 3$ on endothelial cells is needed for angiogenesis [143]. On the other hand, soluble PEX domain inhibits MMP-2 binding to $\alpha v\beta 3$ and blocks angiogenesis [144]. In tumour angiogenesis, MMP-2 and MMP-9 have been shown to be vital for the 'angiogenic switch' when tumours first become vascularized [145,146]. Over-expression of MMP-9 in human breast cancer MCF-7 cells has been reported to result in augmented tumour angiogenesis, tumour growth, and VEGF/VEGF-R2 complex formation, suggesting MMP-9 controls release of VEGF from the extracellular matrix [146]. It has been reported that the implantation of tumour cells into either MMP-2 or MMP-9 knockout mice lead to reduced tumour-induced angiogenesis compared to wild type mice [147,148]. In blood vessels within tumours formed by human neuroblastoma cells implanted in MMP-9 knockout mice, there were decreased numbers of pericytes, suggesting that MMP-9 is needed for recruitment of pericytes to new vessels [149]. Angiogenic factors can induce the expression of MMPs in endothelial cells. Membrane vesicles containing MMP-2, MMP-9, and MT1-MMP can be found in endothelial cells, in some cases confined to invading pseudopodia, and angiogenic stimulation of cells with FGF-2 or VEGF results in shedding of vesicles [150]. It has also been shown that VEGF induces expression of MMP-1 and MMP-2 in endothelial cells, further supporting the role of MMP in angiogenesis [151,152].

It has generally been understood that all MMP inhibitors are inhibitors of angiogenesis, but so far only TIMP-2 has been consistently shown to be a direct inhibitor of angiogenesis [153-155], with varying effects being reported for TIMP-1 and -3 [153,156-158]. Although TIMP-3 has been reported to compete for binding of VEGF to its receptor VEGF-R2 in a genetically modified cell system, the concentration necessary to block VEGF binding is relatively high and has been suggested to be unlikely to play a role in an *in vivo* setting [158]. Nonetheless, TIMP-3-deficient mice show more rapid tumour growth than their wild type litter mates [159].

It is also becoming apparent that MMP and TIMP are a further sub-set of molecules which may be regulated by hypoxia. The balance between MMP/TIMP is likely to influence cell invasion, in the context of angiogenesis (via degradation of extracellular matrix) and/or in terms of invasion by synovium of underlying tissue such cartilage, bone and tendon. A variety of MMP have been shown to be regulated by hypoxia. RA synovial fibroblasts when exposed to hypoxia exhibit increased protein levels of MMP-1 and MMP-3 [160]. In endothelial cells, prolonged hypoxia enhanced MMP-2 (mRNA and protein) expression [161]. In another study using endothelial cells, hypoxia was shown to increase MT1-MMP and MMP-2 mRNA, thereby upregulating MMP-2 activation [162]. Breast cancer cells when cultured in hypoxia also showed increased secretion of MMP-9, with no change in MMP-2 levels [163]. Hypoxia up-regulated gelatinase (MMP-2 and MMP-9) activity in a variety of adenocarcinoma cell lines and increased their invasiveness *in vitro* [164]. Moreover, hypoxic culture of breast cancer cells increased their invasiveness via augmented MT1-MMP expression, as well MMP-2 and/or MMP-9 levels [165,166]. Hypoxia has been reported to upregulate MMP-2 activity and MMP-9 expression in a fibroblast cell line [167], and this has been also reported for other cell types [168,169]. Using DNA micro-arrays, MMP-13 has been identified as a hypoxia-induced gene in carcinoma cells, and postulated to be in part responsible for increased tumour invasiveness [170]. Regulation of TIMP by hypoxia has been reported as well. Hypoxic RA synovial fibroblasts have been show to decrease expression of TIMP-1 at both protein and mRNA levels [160]. In endothelial cells, hypoxia suppressed TIMP-2 mRNA [161]. Conversely, another study showed increased TIMP-2 mRNA in endothelial cells [162]. Trophoblast cells cultured in hypoxia showed increased invasiveness and reduced TIMP-1 protein secretion [163]. Crucially, there is evidence emerging that MT1-MMP is regulated by HIF-2α, thus implicating this key hypoxic pathway in regulation of MMP. This was demonstrated using renal carcinoma cells which lack the von Hippel-Lindau tumour suppressor protein, which targets HIF-α for polyubiquitination by ubiquitin E3 ligases, and in which HIF-α protein is stabilised. In these cells, mRNA for MT1-MMP, MMP-2 and MMP-9, but not for other MMP (MMP-1, MMP-3 and MMP-13), was over-expressed compared to wild-type cells [171,172]. Murine bone marrow derived cells cultured under hypoxic conditions have been found to expand into capillary-like structures via upregulation of MT1-MMP [173]. A functional HRE in the MT1-MMP promoter has been identified [172].

In summary, hypoxia appears to be involved in the regulation of matrix breakdown, via enhanced MMP mRNA and/or protein. There is thus an emerging link between hypoxia, angiogenesis, synovial invasion and disease progression in RA.

Potential for Anti-Angiogenic Therapy in RA

The paradigm for a therapeutic modality in a given disease should take into account underlying pathogenetic mechanisms. In the case of RA, development of TNFα inhibitors resulted from the elucidation of the key role for TNFα in mediating the inflammatory pathways involved in synovitis [174]. In an analogous fashion, the central role of angiogenesis in RA suggests that targeting blood vessels might be beneficial in RA [175,176].

The plethora of angiogenic factors expressed in RA has made it difficult to select the 'ideal' candidate for angiogenesis blockade. Based on the oncology literature, VEGF would be an obvious choice, not only because many research programmes are focusing on development of molecules designed to inhibit VEGF, but also because VEGF is expressed in RA [29,30,38]. As discussed earlier, the effects of VEGF are mediated via binding to two receptors, VEGF-R1 and VEGF-R2, with a soluble form of VEGF-R1 also described [17]. Inhibition of VEGF activity could therefore be achieved using VEGF blocking antibodies to mop up secreted VEGF, soluble VEGF-R1, receptor blocking antibodies to prevent VEGF binding, or interfering with VEGF signalling. Our own studies have utilised human soluble VEGF-R1 in a murine model of arthritis. Acute collagen-induced CIA is induced in genetically susceptible DBA/1 mice using bovine collagen to give a disease which has many similarities with RA, including synovitis, pannus formation, erosion of cartilage and bone, fibrosis and joint rigidity. Treatment is generally subsequent to the onset of visible symptoms (paw swelling and erythema). Soluble VEGF-R1 significantly reduced disease severity in CIA, and the joints appeared grossly normal, with little evidence of joint inflammation or reduced bone and cartilage destruction [177]. Adenovirus expressing soluble VEGF-R1 delivered through a single injection also suppressed both clinical signs of arthritis in mice as well as paw swelling [178]. The effectiveness of VEGF blockade in CIA has also been shown in a number of other studies using anti-VEGF antibodies [179-182]. More recently treatment of murine CIA and the transgenic K/BxN mouse model of arthritis with anti-VEGF-R1 antibody reduced the incidence of disease and suppressed development of clinical symptoms, an effect not observed with anti-VEGF-R2 [183]. The VEGF receptor tyrosine kinase inhibitor PTK787/ZK222584 has also been shown to be effective in arthritis models [184]. At the time of writing, VEGF-targeted treatment such as anti-VEGF antibody or VEGF-R tyrosine kinase inhibitors have not been trialled in RA.

However, the VEGF family of angiogenic molecules encompasses not only VEGF-A, the best studied pro-angiogenic factor and the target for therapies such as bevacizumab, but also placental growth factor (PlGF) [185,186]. Like VEGF, PlGF binds to VEGF-R1 (and soluble VEGF-R1), but, in contrast to VEGF, PlGF does not bind VEGF-R2. PlGF appears not only to induce distinct signalling events via VEGF-R1, but also to amplify VEGF-driven effects through VEGF-R2 and to complex with VEGF/VEGF-R2 forming heterodimeric complexes which transphosphorylate each other [187]. Interestingly, PlGF-deficient mice do not display major vascular abnormalities, unlike mice lacking VEGF [188]. Moreover, loss of PlGF did not impair reproduction, suggesting that this molecule is not essential during physiological angiogenesis, for example during development, skeletal bone growth or in the female reproductive cycle [189]. Instead, PlGF may play a more pronounced role in pathological angiogenesis, as evidenced by impaired tumour growth and vascularisation in mice lacking this molecule. PlGF is also likely to be involved in fracture repair [190]. As a consequence, this is an attractive target to test for potential effectiveness in a disease associated with pathological angiogenesis, such as RA, especially as PlGF is expressed in RA synovial fluid [191]. It is quite possible that the reported effects of soluble VEGF-R1 and VEGF-R1 antibody in CIA are a result of PlGF inhibition, although we failed to find detectable PlGF protein in lysates of arthritic paws [178]. However, in contrast to VEGF blockade, there should be few if any side-effects of PlGF inhibition in terms of disruption of physiological angiogenesis. Moreover, blocking PlGF would avoid the consequences of inhibition of the neuroprotective effects of VEGF, which were highlighted by a study showing that a targeted reduction in VEGF expression in the mouse spinal cord was associated with adult-onset progressive motor neuron degeneration, reminiscent of amyotrophic lateral sclerosis [192]. Although the effect of PlGF blockade in arthritis has not been published, the concept is nonetheless thought-provoking.

Another potential target in RA is the EGF-EGF receptor system, the involvement of which in cancer has motivated interest in these receptors as therapeutic targets. EGF and related molecules such as neuregulins and TGFα bind the ErbB family of receptor tyrosine kinases, comprising ErbB-1 (also known as EGF-R), -2, -3 and -4 [193-195]. This family is the target for several drugs, including trastuzumab (anti-ErbB-2/HER2, approved in metastatic breast cancer for patients whose tumours over-express HER2), cetuximab (anti-EGF-R, approved in advanced metastatic colorectal cancer) and the ErbB tyrosine kinase inhibitor erlotinib (approved for treatment of non-small cell lung cancer). High amounts of EGF in the synovial fluid of RA patients are thought to contribute to the growth of rheumatoid synovial cells [196]. EGF and TGFα have been readily detected in RA synovial membrane and synovial fluid. No expression of EGF-R has been found in the synovium, but there appear to be significant levels of ErbB-2 and EGF receptor ligands such as TGFα and EGF itself in RA synovial membranes [197,198]. EGF stimulated the phosphorylation of ErbB-2 in RA synovial fibroblasts with RA, and this was inhibited by anti-ErbB-2 antibody trastuzumab [197]. Although ErbB inhibitors have not been examined in animal models of arthritis, their link to angiogenesis indicates a viable target in RA therapy.

Rather than blocking the activity of pro-angiogenic growth factors such as VEGF or EGF, an alternative approach would be to utilise known inhibitors of angiogenesis. The best characterised angiogenesis inhibitors are angiostatin and endostatin. Retrovirally-delivered angiostatin has been reported to reduced pannus formation and neovascularisation [199]. Adeno-associated virus expressing angiostatin injected into mouse knee joints before development of arthritis had a comparable disease-modifying effect [200]. Additionally, an HIV vector containing murine angiostatin injected into knee joints after arthritis development also reduced synovial hyperplasia and pannus formation [201]. Angiostatin is itself generated as a result of proteolytic cleavage of plasminogen and comprises the first 4 triple loop disulphide-linked structures of plasminogen, termed kringle domains [202,203]. Urokinase-activated plasmin can also convert plasminogen into a molecule termed K1-5, which contains intact kringles 1-4 and most of the kringle 5 domain. In murine CIA, K1-5 reduced paw swelling and clinical score, as well as joint inflammation and destruction [204]. Endostatin was also found to be effective. Intra-synovial injection of human recombinant endostatin reduced both the volume of grafted synovium and the number of vessels in mice engrafted with human RA synovium [205], whereas systemic administration of endostatin reduced the severity of urine arthritis induced by a cocktail of anti-collagen antibodies [206]. Similarly, endostatin-expressing lentivirus injected directly into the joints of human TNF-transgenic mice before the onset of disease reduced synovial blood vessel density and arthritis severity [207]. These observations with endostatin are of interest, since in human RA patients endostatin levels appear to be similar to non-RA individuals, despite an elevation in VEGF levels, suggesting that angiogenesis in RA occurs as a result of an imbalance between angiogenic stimuli and inhibitors [208].

Other compounds with a wide range of activities which include angiogenesis inhibition have also been studied in arthritis animal models. One such molecule is TNP-470 (AGM-1470), a synthetic analogue of fumagillin, a naturally occurring fungal antibiotic from *Aspergillus fumigatus*. TNP-470 was proven to be effective in the treatment of arthritis in different animal models, including rat CIA and adjuvant-induced arthritis [209-212], and the transgenic K/BxN mouse model of spontaneous arthritis [179]. The mechanism of action was suggested to be inhibition of angiogenesis since peripheral blood VEGF levels were reduced by TNP-470 [211]. In another model of arthritis, in which human RA synovial tissue was implanted into the back of SCID mice, TNP-470 treatment reduced the number of blood vessels and synovial lining cells [213]. Interestingly, TNP-470 and fumagillin both selectively and irreversibly inhibit the enzyme methionine aminopeptidase type 2 isoform, involved in protein maturation. A recent report has described an irreversible inhibitor of methionine aminopeptidase-2 which not only inhibited proliferation of endothelial cells and fibroblasts, but also showed significant *in vivo* activity in a rat model of peptidoglycan-polysaccharide-induced arthritis [214]. This methionine aminopeptidase-2 inhibitor offers a significant advantage over fumagillin and TNP-470, the usefulness of which has been severely limited due to significant neurotoxicity. In contrast, the specific methionine aminopeptidase-2 inhibitor showed a marked reduction in central nervous system side-effects [214].

Further compounds with anti-angiogenic effects include the endogenous oestrogen metabolite 2-methoxyoestradiol, which was shown to be a potent inhibitor of endothelial cell proliferation [215]. Modification of 2-methoxyoestradiol by sulfamate groups yielded a series of compounds with even more potent anti-angiogenic activities [216-218]. While endothelial proliferation and tubule formation were reduced by 2-methoxyoestradiol, this effect was enhanced 10-100-fold by sulfamation, with the effect being mediated most likely through induction of apoptosis [217,219]. Furthermore, 2-methoxyoestradiol bis-sulfamate inhibited tumour growth in nude mice bearing xenografts derived from MDA-MB-435 (oestrogen receptor negative) melanoma cancer cells. At comparable doses 2-methoxyoestradiol did not inhibit tumour growth. Interestingly, the severity of RA is reduced during pregnancy, supporting a possible role for oestrogen in arthritis. The biologically active oestrogen molecule, 17β-oestradiol, has been reported to reduce the severity of murine arthritis [220]. The effect of 2-methoxyoestradiol 2 was also studied in CIA, using murine collagen to induce disease. Treatment with 2-methoxyoestradiol markedly reduced disease and suppressed proliferation of endothelial cells [221], which may be attributable to attenuation of angiogenesis.

CONCLUSIONS

While the impact of TNFα inhibition on management of RA cannot be understated, the incidence of infections means that additional therapies are desirable. Targeting combinations of cytokines, for example IL-1 plus TNFα, is not a viable option, again due to the increased risk of infection. There has been a phenomenal interest in the development of inhibitors of angiogenesis, which represent a major success in the fight against cancer. However while the spotlight has been on malignancies, the many parallels between tumour growth and RA reinforce the hypothesis that inhibition of angiogenesis might also potentially be effective in RA. The concept of angiogenesis inhibition was initially developed for the treatment of cancer in an attempt to starve cancer cells of the nutrients and oxygen they need to grow. Such an approach may also be applied to inflammatory diseases such as RA, whereby invading inflammatory and proliferating cells are starved of nutrients, hindering their ability to exacerbate inflammation. The unique role of VEGF in RA supports the premise that this molecule is an exciting target for therapy in RA. In particular, combined VEGF and TNFα inhibition might avoid the infectious complications seen with other treatment approaches, representing an exciting and emerging area in the treatment of RA.

REFERENCES

[1] Williamson SC, Feldon P. Extensor tendon ruptures in rheumatoid arthritis. *Hand Clin* 1995;11(3):449-59.

[2] Ertel AN. Flexor tendon ruptures in rheumatoid arthritis. *Hand Clin* 1989;5(2):177-90.

[3] Van Doornum S, McColl G, Wicks IP. Accelerated atherosclerosis: an extraarticular feature of rheumatoid arthritis? *Arthritis & Rheumatism* 2002;46(4):862-73.

[4] Gabriel SE, Crowson CS, Kremers HM, Doran MF, Turesson C, O'Fallon WM, et al. Survival in rheumatoid arthritis: a population-based analysis of trends over 40 years. *Arthritis & Rheumatism* 2003;48(1):54-8.

[5] Kaplan MJ. Cardiovascular disease in rheumatoid arthritis. *Curr Opin Rheumatol* 2006;18(3):289-97.

[6] Sihvonen S, Korpela M, Laippala P, Mustonen J, Pasternack A. Death rates and causes of death in patients with rheumatoid arthritis: a population-based study. *Scand J Rheumatol* 2004;33(4):221-227.

[7] Sewell KL, Trentham DE. Pathogenesis of rheumatoid arthritis. *Lancet* 1993;341(8840):283-6.

[8] Brennan FM, Chantry D, Jackson A, Maini R, Feldmann M. Inhibitory effect of TNF alpha antibodies on synovial cell interleukin-1 production in rheumatoid arthritis. *Lancet* 1989;2 (8657)(8657):244-7.

[9] Williams RO, Feldmann M, Maini RN. Anti-tumor necrosis factor ameliorates joint disease in murine collagen-induced arthritis. *Proc Natl Acad Sci U S A* 1992;89(20):9784-8.

[10] Feldmann M, Brennan FM, Maini RN. Role of cytokines in rheumatoid arthritis. *Annu Rev Immunol* 1996;14:397-440.

[11] Maini R, St Clair EW, Breedveld F, Furst D, Kalden J, Weisman M, et al. Infliximab (chimeric anti-tumour necrosis factor alpha monoclonal antibody) versus placebo in rheumatoid arthritis patients receiving concomitant methotrexate: a randomised phase III trial. ATTRACT Study Group. *Lancet* 1999;354(9194):1932-9.

[12] Maini RN, Taylor PC. Anti-cytokine therapy for rheumatoid arthritis. *Annu Rev Med* 2000;51:207-29.

[13] Hochberg MC, Lebwohl MG, Plevy SE, Hobbs KF, Yocum DE. The benefit/risk profile of TNF-blocking agents: findings of a consensus panel. *Semin Arthritis Rheum* 2005;34(6):819-36.

[14] Genovese MC, Cohen S, Moreland L, Lium D, Robbins S, Newmark R, et al. Combination therapy with etanercept and anakinra in the treatment of patients with rheumatoid arthritis who have been treated unsuccessfully with methotrexate. *Arthritis Rheum* 2004;50(5):1412-9.

[15] Folkman J. Angiogenesis. *Annu Rev Med* 2006;57:1-18.

[16] Liekens S, De Clercq E, Neyts J. Angiogenesis: regulators and clinical applications. *Biochem Pharmacol* 2001;61(3):253-70.

[17] Ferrara N, Gerber HP, LeCouter J. The biology of VEGF and its receptors. *Nat Med* 2003;9(6):669-76.

[18] Olsson AK, Dimberg A, Kreuger J, Claesson-Welsh L. VEGF receptor signalling - in control of vascular function. *Nat Rev Mol Cell Biol* 2006;7(5):359-71.

[19] Roy H, Bhardwaj S, Yla-Herttuala S. Biology of vascular endothelial growth factors. *FEBS Lett* 2006;580(12):2879-87.

[20] Eskens FA. Angiogenesis inhibitors in clinical development; where are we now and where are we going? *Br J Cancer* 2004;90(1):1-7.

[21] Kabbinavar F, Hurwitz HI, Fehrenbacher L, Meropol NJ, Novotny WF, Lieberman G, et al. Phase II, randomized trial comparing bevacizumab plus fluorouracil (FU)/leucovorin (LV) with FU/LV alone in patients with metastatic colorectal cancer. *J Clin Oncol* 2003;21(1):60-5.

[22] Brown RA, Weiss JB, Tomlinson IW, Phillips P, Kumar S. Angiogenic factor from synovial fluid resembling that from tumours. *Lancet* 1980;1(8170):682-5.

[23] Semble EL, Turner RA, McCrickard EL. Rheumatoid arthritis and osteoarthritis synovial fluid effects on primary human endothelial cell cultures. *J Rheumatol* 1985;12(2):237-41.

[24] Kumar P, Erroi A, Sattar A, Kumar S. Weibel-Palade bodies as a marker for neovascularization induced by tumor and rheumatoid angiogenesis factors. *Cancer Res* 1985;45(9):4339-48.

[25] Fearon U, Griosios K, Fraser A, Reece R, Emery P, Jones PF, et al. Angiopoietins, growth factors, and vascular morphology in early arthritis. *J Rheumatol* 2003;30(2):260-8.

[26] Ceponis A, Konttinen YT, Imai S, Tamulaitiene M, Li TF, Xu JW, et al. Synovial lining, endothelial and inflammatory mononuclear cell proliferation in synovial membranes in psoriatic and reactive arthritis: a comparative quantitative morphometric study. *Br. J. Rheumatol.* 1998;37(2):170-8.

[27] Walsh DA, Wade M, Mapp PI, Blake DR. Focally regulated endothelial proliferation and cell death in human synovium. *Am J Pathol* 1998;152(3):691-702.

[28] Fraser A, Fearon U, Reece R, Emery P, Veale DJ. Matrix metalloproteinase 9, apoptosis, and vascular morphology in early arthritis. *Arthritis Rheum* 2001;44(9):2024-8.

[29] Koch AE, Harlow LA, Haines GK, Amento EP, Unemori EN, Wong WL, et al. Vascular endothelial growth factor. A cytokine modulating endothelial function in rheumatoid arthritis. *J Immunol* 1994;152(8):4149-56.

[30] Fava RA, Olsen NJ, Spencer-Green G, Yeo KT, Yeo TK, Berse B, et al. Vascular permeability factor/endothelial growth factor (VPF/VEGF): accumulation and expression in human synovial fluids and rheumatoid synovial tissue. *J Exp Med* 1994;180(1):341-6.

[31] Vignola S, Picco P, Falcini F, Sabatini F, Buoncompagni A, Gattorno M. Serum and synovial fluid concentration of vascular endothelial growth factor in juvenile idiopathic arthritides. *Rheumatology (Oxford)* 2002;41(6):691-6.

[32] Maeno N, Takei S, Imanaka H, Takasaki I, Kitajima I, Maruyama I, et al. Increased circulating vascular endothelial growth factor is correlated with disease activity in polyarticular juvenile rheumatoid arthritis. *J Rheumatol* 1999;26(10):2244-8.

[33] Kikuchi K, Kubo M, Kadono T, Yazawa N, Ihn H, Tamaki K. Serum concentrations of vascular endothelial growth factor in collagen diseases. *Br J Dermatol* 1998;139(6):1049-51.

[34] Goldberger C, Dulak J, Duftner C, Weidinger F, Falkenbach A, Schirmer M. Vascular endothelial growth factor (VEGF) in ankylosing spondylitis--a pilot study. *Wien Med Wochenschr* 2002;152(9-10):223-5.

[35] Drouart M, Saas P, Billot M, Cedoz JP, Tiberghien P, Wendling D, et al. High serum vascular endothelial growth factor correlates with disease activity of spondylarthropathies. *Clin Exp Immunol* 2003;132(1):158-62.

[36] Ballara S, Taylor PC, Reusch P, Marme D, Feldmann M, Maini RN, et al. Raised serum vascular endothelial growth factor levels are associated with destructive change in inflammatory arthritis. *Arthritis Rheum* 2001;44(9):2055-64.

[37] Harada M, Mitsuyama K, Yoshida H, Sakisaka S, Taniguchi E, Kawaguchi T, et al. Vascular endothelial growth factor in patients with rheumatoid arthritis. *Scand J Rheumatol* 1998;27(5):377-80.

[38] Paleolog EM, Young S, Stark AC, McCloskey RV, Feldmann M, Maini RN. Modulation of angiogenic vascular endothelial growth factor by tumor necrosis factor alpha and interleukin-1 in rheumatoid arthritis. *Arthritis Rheum* 1998;41(7):1258-65.

[39] Lee SS, Joo YS, Kim WU, Min DJ, Min JK, Park SH, et al. Vascular endothelial growth factor levels in the serum and synovial fluid of patients with rheumatoid arthritis. *Clin Exp Rheumatol* 2001;19(3):321-4.

[40] Sone H, Sakauchi M, Takahashi A, Suzuki H, Inoue N, Iida K, et al. Elevated levels of vascular endothelial growth factor in the sera of patients with rheumatoid arthritis correlation with disease activity. *Life Sci* 2001;69(16):1861-9.

[41] Latour F, Zabraniecki L, Dromer C, Brouchet A, Durroux R, Fournie B. Does vascular endothelial growth factor in the rheumatoid synovium predict joint destruction? A clinical, radiological, and pathological study in 12 patients monitored for 10 years. *Joint Bone Spine* 2001;68(6):493-8.

[42] Nakahara H, Song J, Sugimoto M, Hagihara K, Kishimoto T, Yoshizaki K, et al. Anti-interleukin-6 receptor antibody therapy reduces vascular endothelial growth factor production in rheumatoid arthritis. *Arthritis Rheum* 2003;48(6):1521-9.

[43] Klimiuk PA, Sierakowski S, Domyslawska I, Fiedorczyk M, Chwiecko J. Reduction of soluble adhesion molecules (sICAM-1, sVCAM-1, and sE-selectin) and vascular endothelial growth factor levels in serum of rheumatoid arthritis patients following multiple intravenous infusions of infliximab. *Arch Immunol Ther Exp (Warsz)* 2004;52(1):36-42.

[44] Aggarwal A, Panda S, Misra R. Effect of etanercept on matrix metalloproteinases and angiogenic vascular endothelial growth factor: a time kinetic study. *Ann Rheum Dis* 2004;63(7):891-2.

[45] Macias I, Garcia-Perez S, Ruiz-Tudela M, Medina F, Chozas N, Giron-Gonzalez JA. Modification of pro- and antiinflammatory cytokines and vascular-related molecules by tumor necrosis factor-a blockade in patients with rheumatoid arthritis. *J Rheumatol* 2005;32(11):2102-8.

[46] Nagashima M, Wauke K, Hirano D, Ishigami S, Aono H, Takai M, et al. Effects of combinations of anti-rheumatic drugs on the production of vascular endothelial growth factor and basic fibroblast growth factor in cultured synoviocytes and patients with rheumatoid arthritis. *Rheumatology (Oxford)* 2000;39(11):1255-62.

[47] Klimiuk PA, Sierakowski S, Latosiewicz R, Cylwik JP, Cylwik B, Skowronski J, et al. Soluble adhesion molecules (ICAM-1, VCAM-1, and E-selectin) and vascular endothelial growth factor (VEGF) in patients with distinct variants of rheumatoid synovitis. *Ann Rheum Dis* 2002;61(9):804-9.

[48] Kuryliszyn-Moskal A, Klimiuk PA, Sierakowski S, Ciolkiewicz M. A study on vascular endothelial growth factor and endothelin-1 in patients with extra-articular involvement of rheumatoid arthritis. *Clin Rheumatol* 2006;25(3):314-9.

[49] Nagashima M, Yoshino S, Ishiwata T, Asano G. Role of vascular endothelial growth factor in angiogenesis of rheumatoid arthritis. *J Rheumatol* 1995;22(9):1624-30.

[50] Kasama T, Kobayashi K, Yajima N, Shiozawa F, Yoda Y, Takeuchi HT, et al. Expression of vascular endothelial growth factor by synovial fluid neutrophils in rheumatoid arthritis (RA). *Clin Exp Immunol* 2000;121(3):533-8.

[51] Taichman NS, Young S, Cruchley AT, Taylor P, Paleolog E. Human neutrophils secrete vascular endothelial growth factor. *Journal of Leukocyte Biology* 1997;62(3):397-400.

[52] Ben-Av P, Crofford LJ, Wilder RL, Hla T. Induction of vascular endothelial growth factor expression in synovial fibroblasts by prostaglandin E and interleukin-1: a potential mechanism for inflammatory angiogenesis. *FEBS Lett* 1995;372(1):83-7.

[53] Jackson JR, Minton JA, Ho ML, Wei N, Winkler JD. Expression of vascular endothelial growth factor in synovial fibroblasts is induced by hypoxia and interleukin 1beta. *J Rheumatol* 1997;24(7):1253-9.

[54] Bottomley MJ, Webb NJ, Watson CJ, Holt PJ, Freemont AJ, Brenchley PE. Peripheral blood mononuclear cells from patients with rheumatoid arthritis spontaneously secrete vascular endothelial growth factor (VEGF): specific up-regulation by tumour necrosis factor-alpha (TNF-alpha) in synovial fluid. *Clin Exp Immunol* 1999;117(1):171-6.

[55] Berse B, Hunt JA, Diegel RJ, Morganelli P, Yeo K, Brown F, et al. Hypoxia augments cytokine (transforming growth factor-beta (TGF-beta) and IL-1)-induced vascular

endothelial growth factor secretion by human synovial fibroblasts. *Clin Exp Immunol* 1999;115(1):176-82.

[56] Jain A, Nanchahal J, Troeberg L, Green P, Brennan F. Production of cytokines, vascular endothelial growth factor, matrix metalloproteinases, and tissue inhibitor of metalloproteinases 1 by tenosynovium demonstrates its potential for tendon destruction in rheumatoid arthritis. *Arthritis Rheum* 2001;44(8):1754-60.

[57] Pufe T, Petersen W, Tillmann B, Mentlein R. Splice variants VEGF121 and VEGF165 of the angiogenic peptide vascular endothelial cell growth factor are expressed in the synovial tissue of patients with rheumatoid arthritis. *J Rheumatol* 2001;28(7):1482-5.

[58] Giatromanolaki A, Sivridis E, Athanassou N, Zois E, Thorpe PE, Brekken RA, et al. The angiogenic pathway "vascular endothelial growth factor/flk-1(KDR)-receptor" in rheumatoid arthritis and osteoarthritis. *J Pathol* 2001;194(1):101-8.

[59] Ikeda M, Hosoda Y, Hirose S, Okada Y, Ikeda E. Expression of vascular endothelial growth factor isoforms and their receptors Flt-1, KDR, and neuropilin-1 in synovial tissues of rheumatoid arthritis. *J Pathol* 2000;191(4):426-33.

[60] Asahara T, Takahashi T, Masuda H, Kalka C, Chen D, Iwaguro H, et al. VEGF contributes to postnatal neovascularization by mobilizing bone marrow-derived endothelial progenitor cells. *Embo J* 1999;18(14):3964-72.

[61] Kalka C, Masuda H, Takahashi T, Gordon R, Tepper O, Gravereaux E, et al. Vascular endothelial growth factor(165) gene transfer augments circulating endothelial progenitor cells in human subjects. *Circ Res* 2000;86(12):1198-202.

[62] Ruger B, Giurea A, Wanivenhaus AH, Zehetgruber H, Hollemann D, Yanagida G, et al. Endothelial precursor cells in the synovial tissue of patients with rheumatoid arthritis and osteoarthritis. *Arthritis Rheum* 2004;50(7):2157-66.

[63] Hirohata S, Yanagida T, Nampei A, Kunugiza Y, Hashimoto H, Tomita T, et al. Enhanced generation of endothelial cells from CD34+ cells of the bone marrow in rheumatoid arthritis: possible role in synovial neovascularization. *Arthritis Rheum* 2004;50(12):3888-96.

[64] Grisar J, Aletaha D, Steiner CW, Kapral T, Steiner S, Seidinger D, et al. Depletion of endothelial progenitor cells in the peripheral blood of patients with rheumatoid arthritis. *Circulation* 2005;111(2):204-11.

[65] Scott BB, Zaratin PF, Colombo A, Hansbury MJ, Winkler JD, Jackson JR. Constitutive expression of angiopoietin-1 and -2 and modulation of their expression by inflammatory cytokines in rheumatoid arthritis synovial fibroblasts. *J Rheumatol* 2002;29(2):230-9.

[66] Gravallese EM, Pettit AR, Lee R, Madore R, Manning C, Tsay A, et al. Angiopoietin-1 is expressed in the synovium of patients with rheumatoid arthritis and is induced by tumour necrosis factor alpha. *Ann Rheum Dis* 2003;62(2):100-7.

[67] de Busk LM, Chen Y, Nishishita T, Chen J, Thomas JW, Lin PC. Tie2 receptor tyrosine kinase, a major mediator of tumor necrosis factor alpha-induced angiogenesis in rheumatoid arthritis. *Arthritis Rheum* 2003;48(9):2461-71.

[68] Shahrara S, Volin MV, Connors MA, Haines GK, Koch AE. Differential expression of the angiogenic Tie receptor family in arthritic and normal synovial tissue. *Arthritis Res* 2002;4(3):201-8.

[69] Salvador G, Sanmarti R, Gil-Torregrosa B, Garcia-Peiro A, Rodriguez-Cros JR, Canete JD. Synovial vascular patterns and angiogenic factors expression in synovial tissue and serum of patients with rheumatoid arthritis. *Rheumatology (Oxford)* 2006;45(8):966-71.

[70] Sano H, Forough R, Maier JA, Case JP, Jackson A, Engleka K, et al. Detection of high levels of heparin binding growth factor-1 (acidic fibroblast growth factor) in inflammatory arthritic joints. *J Cell Biol* 1990;110(4):1417-26.

[71] Sano H, Engleka K, Mathern P, Hla T, Crofford LJ, Remmers EF, et al. Coexpression of phosphotyrosine-containing proteins, platelet-derived growth factor-B, and fibroblast growth factor-1 in situ in synovial tissues of patients with rheumatoid arthritis and Lewis rats with adjuvant or streptococcal cell wall arthritis. *J. Clin. Invest.* 1993;91(2):553-65.

[72] Nakashima M, Eguchi K, Aoyagi T, Yamashita I, Ida H, Sakai M, et al. Expression of basic fibroblast growth factor in synovial tissues from patients with rheumatoid arthritis: detection by immunohistological staining and in situ hybridisation. *Ann Rheum Dis* 1994;53(1):45-50.

[73] Remmers EF, Sano H, Lafyatis R, Case JP, Kumkumian GK, Hla T, et al. Production of platelet derived growth factor B chain (PDGF-B/c-sis) mRNA and immunoreactive PDGF B-like polypeptide by rheumatoid synovium: coexpression with heparin binding acidic fibroblast growth factor-1. *J. Rheumatol.* 1991;18(1):7-13.

[74] Koch AE, Halloran MM, Hosaka S, Shah MR, Haskell CJ, Baker SK, et al. Hepatocyte growth factor. A cytokine mediating endothelial migration in inflammatory arthritis. *Arthritis & Rheumatism* 1996;39(9):1566-75.

[75] Fava RA, Hunt JA, Yeo KT, Brown F, Berse B. Induction of vascular permeability factor (vascular endothelial growth factor) by TGFb in synovial fibroblasts. *Arthritis & Rheumatism* 1995;38:S343.

[76] Blake DR, Merry P, Unsworth J, Kidd BL, Outhwaite JM, Ballard R, et al. Hypoxic-reperfusion injury in the inflamed human joint. *Lancet* 1989;1(8633):289-93.

[77] Lund-Olesen K. Oxygen tension in synovial fluids. Arthritis Rheum 1970;13(6):769-76.

[78] Etherington PJ, Winlove P, Taylor P, Paleolog E, Miotla J. VEGF release is associated with reduced oxygen tensions in experimental inflammatory arthritis. *Clinical and Experimental Rheumatology* 2002;20(6):799-805.

[79] Stevens CR, Blake DR, Merry P, Revell PA, Levick JR. A comparative study by morphometry of the microvasculature in normal and rheumatoid synovium. *Arthritis Rheum* 1991;34(12):1508-13.

[80] Naughton D, Whelan M, Smith EC, Williams R, Blake DR, Grootveld M. An investigation of the abnormal metabolic status of synovial fluid from patients with rheumatoid arthritis by high field proton nuclear magnetic resonance spectroscopy. *FEBS Lett* 1993;317(1-2):135-8.

[81] Merry P, Grootveld M, Blake DR. Hypoxic-reperfusion injury in inflamed joints. *Lancet* 1989;1(8645):1023.

[82] Naughton DP. Hypoxia-induced upregulation of the glycolytic enzyme glucose-6-phosphate isomerase perpetuates rheumatoid arthritis. *Med Hypotheses* 2003;60(3):332-4.

[83] Semenza GL. HIF-1, O(2), and the 3 PHDs: how animal cells signal hypoxia to the nucleus. *Cell* 2001;107(1):1-3.

[84] Masson N, Ratcliffe PJ. HIF prolyl and asparaginyl hydroxylases in the biological response to intracellular O2 levels. *J Cell Sci* 2003;116(Pt 15):3041-9.

[85] Jaakkola P, Mole DR, Tian YM, Wilson MI, Gielbert J, Gaskell SJ, et al. Targeting of HIF-alpha to the von Hippel-Lindau ubiquitylation complex by O2-regulated prolyl hydroxylation. *Science* 2001;292(5516):468-72.

[86] Lando D, Gorman JJ, Whitelaw ML, Peet DJ. Oxygen-dependent regulation of hypoxia-inducible factors by prolyl and asparaginyl hydroxylation. *Eur J Biochem* 2003;270(5):781-90.

[87] Brahimi-Horn C, Mazure N, Pouyssegur J. Signalling via the hypoxia-inducible factor-1alpha requires multiple posttranslational modifications. *Cell Signal* 2005;17(1):1-9.

[88] Kallio PJ, Okamoto K, O'Brien S, Carrero P, Makino Y, Tanaka H, et al. Signal transduction in hypoxic cells: inducible nuclear translocation and recruitment of the CBP/p300 coactivator by the hypoxia-inducible factor-1alpha. *Embo J* 1998;17(22):6573-86.

[89] Gaber T, Dziurla R, Tripmacher R, Burmester GR, Buttgereit F. Hypoxia inducible factor (HIF) in rheumatology: low O2! See what HIF can do! *Ann Rheum Dis* 2005;64(7):971-80.

[90] Jain A, Kiriakidis S, Brennan F, Sandison A, Paleolog E, Nanchahal J. Targeting rheumatoid tenosynovial angiogenesis with cytokine inhibitors. *Clin Orthop Relat Res* 2006;446:268-77.

[91] Hitchon C, Wong K, Ma G, Reed J, Lyttle D, El-Gabalawy H. Hypoxia-induced production of stromal cell-derived factor 1 (CXCL12) and vascular endothelial growth factor by synovial fibroblasts. *Arthritis Rheum* 2002;46(10):2587-97.

[92] Hollander AP, Corke KP, Freemont AJ, Lewis CE. Expression of hypoxia-inducible factor 1alpha by macrophages in the rheumatoid synovium: implications for targeting of therapeutic genes to the inflamed joint. *Arthritis Rheum* 2001;44(7):1540-4.

[93] Giatromanolaki A, Sivridis E, Maltezos E, Athanassou N, Papazoglou D, Gatter KC, et al. Upregulated hypoxia inducible factor-1alpha and -2alpha pathway in rheumatoid arthritis and osteoarthritis. *Arthritis Res Ther* 2003;5(4):R193-201.

[94] Peters CL, Morris CJ, Mapp PI, Blake DR, Lewis CE, Winrow VR. The transcription factors hypoxia-inducible factor 1alpha and Ets-1 colocalize in the hypoxic synovium of inflamed joints in adjuvant-induced arthritis. *Arthritis Rheum* 2004;50(1):291-6.

[95] Sivakumar B, Akhavani M, Kang N, Taylor P, Paleolog E. Hypoxia-driven angiogenesis is a key feature of tendon disease in rheumatoid arthritis. *Rheumatology (Oxford)* 2006;45(Supplement 1):i39 (41).

[96] Cramer T, Yamanishi Y, Clausen BE, Forster I, Pawlinski R, Mackman N, et al. HIF-1alpha is essential for myeloid cell-mediated inflammation. *Cell* 2003;112(5):645-57.

[97] Ceradini DJ, Kulkarni AR, Callaghan MJ, Tepper OM, Bastidas N, Kleinman ME, et al. Progenitor cell trafficking is regulated by hypoxic gradients through HIF-1 induction of SDF-1. *Nat Med* 2004;10(8):858-64.

[98] Moore MA, Hattori K, Heissig B, Shieh JH, Dias S, Crystal RG, et al. Mobilization of endothelial and hematopoietic stem and progenitor cells by adenovector-mediated

elevation of serum levels of SDF-1, VEGF, and angiopoietin-1. *Ann N Y Acad Sci* 2001;938:36-45; discussion 45-7.

[99] Woolley DE, Tetlow LC. Observations on the microenvironmental nature of cartilage degradation in rheumatoid arthritis. *Ann Rheum Dis* 1997;56(3):151-61.

[100] Dalton S, Cawston TE, Riley GP, Bayley IJ, Hazleman BL. Human shoulder tendon biopsy samples in organ culture produce procollagenase and tissue inhibitor of metalloproteinases. *Ann Rheum Dis* 1995;54(7):571-7.

[101] Nagase H, Woessner JF, Jr. Matrix metalloproteinases. *J Biol Chem* 1999;274(31):21491-4.

[102] Woessner JF, Jr. MMPs and TIMPs. An historical perspective. *Methods Mol Biol* 2001;151:1-23.

[103] McCawley LJ, Matrisian LM. Matrix metalloproteinases: they're not just for matrix anymore! *Curr Opin Cell Biol* 2001;13(5):534-40.

[104] Martel-Pelletier J, McCollum R, Fujimoto N, Obata K, Cloutier JM, Pelletier JP. Excess of metalloproteases over tissue inhibitor of metalloprotease may contribute to cartilage degradation in osteoarthritis and rheumatoid arthritis. *Lab Invest* 1994;70(6):807-15.

[105] Clark IM, Powell LK, Ramsey S, Hazleman BL, Cawston TE. The measurement of collagenase, tissue inhibitor of metalloproteinases (TIMP), and collagenase-TIMP complex in synovial fluids from patients with osteoarthritis and rheumatoid arthritis. *Arthritis Rheum* 1993;36(3):372-9.

[106] Yoshihara Y, Nakamura H, Obata K, Yamada H, Hayakawa T, Fujikawa K, et al. Matrix metalloproteinases and tissue inhibitors of metalloproteinases in synovial fluids from patients with rheumatoid arthritis or osteoarthritis. *Ann Rheum Dis* 2000;59(6):455-61.

[107] Konttinen YT, Ainola M, Valleala H, Ma J, Ida H, Mandelin J, et al. Analysis of 16 different matrix metalloproteinases (MMP-1 to MMP-20) in the synovial membrane: different profiles in trauma and rheumatoid arthritis. *Ann Rheum Dis* 1999;58(11):691-7.

[108] Hanemaaijer R, Sorsa T, Konttinen YT, Ding Y, Sutinen M, Visser H, et al. Matrix metalloproteinase-8 is expressed in rheumatoid synovial fibroblasts and endothelial cells. Regulation by tumor necrosis factor-alpha and doxycycline. *J Biol Chem* 1997;272(50):31504-9.

[109] Azzo W, Woessner JF, Jr. Purification and characterization of an acid metalloproteinase from human articular cartilage. *J Biol Chem* 1986;261(12):5434-41.

[110] Yu Q, Stamenkovic I. Localization of matrix metalloproteinase 9 to the cell surface provides a mechanism for CD44-mediated tumor invasion. *Genes Dev* 1999;13(1):35-48.

[111] Visse R, Nagase H. Matrix metalloproteinases and tissue inhibitors of metalloproteinases: structure, function, and biochemistry. *Circ Res* 2003;92(8):827-39.

[112] Gruber BL, Sorbi D, French DL, Marchese MJ, Nuovo GJ, Kew RR, et al. Markedly elevated serum MMP-9 (gelatinase B) levels in rheumatoid arthritis: a potentially useful laboratory marker. *Clin Immunol Immunopathol* 1996;78(2):161-71.

[113] Zhu P, Lu N, Shi ZG, Zhou J, Wu ZB, Yang Y, et al. CD147 overexpression on synoviocytes in rheumatoid arthritis enhances matrix metalloproteinase production and invasiveness of synoviocytes. *Arthritis Res Ther* 2006;8(2):R44.

[114] Jain A, Brennan F, Troeberg L, Nanchahal J. The role of matrix metalloproteinases in rheumatoid tendon disease. *J Hand Surg [Am]* 2002;27(6):1059-64.

[115] Sato H, Takino T, Okada Y, Cao J, Shinagawa A, Yamamoto E, et al. A matrix metalloproteinase expressed on the surface of invasive tumour cells. *Nature* 1994;370(6484):61-5.

[116] Itoh Y, Kajita M, Kinoh H, Mori H, Okada A, Seiki M. Membrane type 4 matrix metalloproteinase (MT4-MMP, MMP-17) is a glycosylphosphatidylinositol-anchored proteinase. *J Biol Chem* 1999;274(48):34260-6.

[117] Udayakumar TS, Chen ML, Bair EL, Von Bredow DC, Cress AE, Nagle RB, et al. Membrane type-1-matrix metalloproteinase expressed by prostate carcinoma cells cleaves human laminin-5 beta3 chain and induces cell migration. *Cancer Res* 2003;63(9):2292-9.

[118] Seiki M. Membrane-type 1 matrix metalloproteinase: a key enzyme for tumor invasion. *Cancer Lett* 2003;194(1):1-11.

[119] Holmbeck K, Bianco P, Caterina J, Yamada S, Kromer M, Kuznetsov SA, et al. MT1-MMP-deficient mice develop dwarfism, osteopenia, arthritis, and connective tissue disease due to inadequate collagen turnover. *Cell* 1999;99(1):81-92.

[120] Holmbeck K, Bianco P, Pidoux I, Inoue S, Billinghurst RC, Wu W, et al. The metalloproteinase MT1-MMP is required for normal development and maintenance of osteocyte processes in bone. *J Cell Sci* 2005;118(Pt 1):147-56.

[121] Brew K, Dinakarpandian D, Nagase H. Tissue inhibitors of metalloproteinases: evolution, structure and function. *Biochim Biophys Acta* 2000;1477(1-2):267-83.

[122] Crocker SJ, Pagenstecher A, Campbell IL. The TIMPs tango with MMPs and more in the central nervous system. *J Neurosci Res* 2004;75(1):1-11.

[123] Lambert E, Dasse E, Haye B, Petitfrere E. TIMPs as multifacial proteins. *Crit Rev Oncol Hematol* 2004;49(3):187-98.

[124] Gomez DE, Alonso DF, Yoshiji H, Thorgeirsson UP. Tissue inhibitors of metalloproteinases: structure, regulation and biological functions. *Eur J Cell Biol* 1997;74(2):111-22.

[125] Wilhelm SM, Collier IE, Marmer BL, Eisen AZ, Grant GA, Goldberg GI. SV40-transformed human lung fibroblasts secrete a 92-kDa type IV collagenase which is identical to that secreted by normal human macrophages. *J Biol Chem* 1989;264(29):17213-21.

[126] Howard EW, Banda MJ. Binding of tissue inhibitor of metalloproteinases 2 to two distinct sites on human 72-kDa gelatinase. Identification of a stabilization site. *J Biol Chem* 1991;266(27):17972-7.

[127] Butler GS, Apte SS, Willenbrock F, Murphy G. Human tissue inhibitor of metalloproteinases 3 interacts with both the N- and C-terminal domains of gelatinases A and B. Regulation by polyanions. *J Biol Chem* 1999;274(16):10846-51.

[128] Bigg HF, Shi YE, Liu YE, Steffensen B, Overall CM. Specific, high affinity binding of tissue inhibitor of metalloproteinases-4 (TIMP-4) to the COOH-terminal hemopexin-

like domain of human gelatinase A. TIMP-4 binds progelatinase A and the COOH-terminal domain in a similar manner to TIMP-2. *J Biol Chem* 1997;272(24):15496-500.

[129] van Hinsbergh VW, Engelse MA, Quax PH. Pericellular proteases in angiogenesis and vasculogenesis. *Arterioscler Thromb Vasc Biol* 2006;26(4):716-28.

[130] Zhou Z, Apte SS, Soininen R, Cao R, Baaklini GY, Rauser RW, et al. Impaired endochondral ossification and angiogenesis in mice deficient in membrane-type matrix metalloproteinase I. *Proc Natl Acad Sci U S A* 2000;97(8):4052-7.

[131] Sounni NE, Devy L, Hajitou A, Frankenne F, Munaut C, Gilles C, et al. MT1-MMP expression promotes tumor growth and angiogenesis through an up-regulation of vascular endothelial growth factor expression. *Faseb J* 2002;16(6):555-64.

[132] Sounni NE, Roghi C, Chabottaux V, Janssen M, Munaut C, Maquoi E, et al. Up-regulation of vascular endothelial growth factor-A by active membrane-type 1 matrix metalloproteinase through activation of Src-tyrosine kinases. *J Biol Chem* 2004;279(14):13564-74.

[133] Jeong JW, Cha HJ, Yu DY, Seiki M, Kim KW. Induction of membrane-type matrix metalloproteinase-1 stimulates angiogenic activities of bovine aortic endothelial cells. *Angiogenesis* 1999;3(2):167-74.

[134] Hiraoka N, Allen E, Apel IJ, Gyetko MR, Weiss SJ. Matrix metalloproteinases regulate neovascularization by acting as pericellular fibrinolysins. *Cell* 1998;95(3):365-77.

[135] Koike T, Vernon RB, Hamner MA, Sadoun E, Reed MJ. MT1-MMP, but not secreted MMPs, influences the migration of human microvascular endothelial cells in 3-dimensional collagen gels. *J Cell Biochem* 2002;86(4):748-58.

[136] Lafleur MA, Handsley MM, Knauper V, Murphy G, Edwards DR. Endothelial tubulogenesis within fibrin gels specifically requires the activity of membrane-type-matrix metalloproteinases (MT-MMPs). *J Cell Sci* 2002;115(Pt 17):3427-38.

[137] Collen A, Hanemaaijer R, Lupu F, Quax PH, van Lent N, Grimbergen J, et al. Membrane-type matrix metalloproteinase-mediated angiogenesis in a fibrin-collagen matrix. *Blood* 2003;101(5):1810-7.

[138] Chun TH, Sabeh F, Ota I, Murphy H, McDonagh KT, Holmbeck K, et al. MT1-MMP-dependent neovessel formation within the confines of the three-dimensional extracellular matrix. *J Cell Biol* 2004;167(4):757-67.

[139] Fisher C, Gilbertson-Beadling S, Powers EA, Petzold G, Poorman R, Mitchell MA. Interstitial collagenase is required for angiogenesis in vitro. *Dev Biol* 1994;162(2):499-510.

[140] Huo N, Ichikawa Y, Kamiyama M, Ishikawa T, Hamaguchi Y, Hasegawa S, et al. MMP-7 (matrilysin) accelerated growth of human umbilical vein endothelial cells. *Cancer Lett* 2002;177(1):95-100.

[141] Nishizuka I, Ichikawa Y, Ishikawa T, Kamiyama M, Hasegawa S, Momiyama N, et al. Matrilysin stimulates DNA synthesis of cultured vascular endothelial cells and induces angiogenesis in vivo. *Cancer Lett* 2001;173(2):175-82.

[142] Xu J, Rodriguez D, Petitclerc E, Kim JJ, Hangai M, Moon YS, et al. Proteolytic exposure of a cryptic site within collagen type IV is required for angiogenesis and tumor growth in vivo. *J Cell Biol* 2001;154(5):1069-79.

[143] Silletti S, Kessler T, Goldberg J, Boger DL, Cheresh DA. Disruption of matrix metalloproteinase 2 binding to integrin alpha vbeta 3 by an organic molecule inhibits angiogenesis and tumor growth in vivo. *Proc Natl Acad Sci U S A* 2001;98(1):119-24.

[144] Brooks PC, Silletti S, von Schalscha TL, Friedlander M, Cheresh DA. Disruption of angiogenesis by PEX, a noncatalytic metalloproteinase fragment with integrin binding activity. *Cell* 1998;92(3):391-400.

[145] Hanahan D, Folkman J. Patterns and emerging mechanisms of the angiogenic switch during tumorigenesis. *Cell* 1996;86(3):353-64.

[146] Bergers G, Brekken R, McMahon G, Vu TH, Itoh T, Tamaki K, et al. Matrix metalloproteinase-9 triggers the angiogenic switch during carcinogenesis. *Nat Cell Biol* 2000;2(10):737-44.

[147] Itoh T, Tanioka M, Yoshida H, Yoshioka T, Nishimoto H, Itohara S. Reduced angiogenesis and tumor progression in gelatinase A-deficient mice. *Cancer Res* 1998;58(5):1048-51.

[148] Fang J, Shing Y, Wiederschain D, Yan L, Butterfield C, Jackson G, et al. Matrix metalloproteinase-2 is required for the switch to the angiogenic phenotype in a tumor model. *Proc Natl Acad Sci U S A* 2000;97(8):3884-9.

[149] Chantrain CF, Shimada H, Jodele S, Groshen S, Ye W, Shalinsky DR, et al. Stromal matrix metalloproteinase-9 regulates the vascular architecture in neuroblastoma by promoting pericyte recruitment. *Cancer Res* 2004;64(5):1675-86.

[150] Nguyen M, Arkell J, Jackson CJ. Human endothelial gelatinases and angiogenesis. *Int J Biochem Cell Biol* 2001;33(10):960-70.

[151] Unemori EN, Ferrara N, Bauer EA, Amento EP. Vascular endothelial growth factor induces interstitial collagenase expression in human endothelial cells. *J Cell Physiol* 1992;153(3):557-62.

[152] Lamoreaux WJ, Fitzgerald ME, Reiner A, Hasty KA, Charles ST. Vascular endothelial growth factor increases release of gelatinase A and decreases release of tissue inhibitor of metalloproteinases by microvascular endothelial cells in vitro. *Microvasc Res* 1998;55(1):29-42.

[153] Murphy AN, Unsworth EJ, Stetler-Stevenson WG. Tissue inhibitor of metalloproteinases-2 inhibits bFGF-induced human microvascular endothelial cell proliferation. *J Cell Physiol* 1993;157(2):351-8.

[154] Li H, Lindenmeyer F, Grenet C, Opolon P, Menashi S, Soria C, et al. AdTIMP-2 inhibits tumor growth, angiogenesis, and metastasis, and prolongs survival in mice. *Hum Gene Ther* 2001;12(5):515-26.

[155] Fernandez CA, Butterfield C, Jackson G, Moses MA. Structural and functional uncoupling of the enzymatic and angiogenic inhibitory activities of tissue inhibitor of metalloproteinase-2 (TIMP-2): loop 6 is a novel angiogenesis inhibitor. *J Biol Chem* 2003;278(42):40989-95.

[156] Hayakawa T, Yamashita K, Tanzawa K, Uchijima E, Iwata K. Growth-promoting activity of tissue inhibitor of metalloproteinases-1 (TIMP-1) for a wide range of cells. A possible new growth factor in serum. *FEBS Lett* 1992;298(1):29-32.

[157] Anand-Apte B, Pepper MS, Voest E, Montesano R, Olsen B, Murphy G, et al. Inhibition of angiogenesis by tissue inhibitor of metalloproteinase-3. *Invest Ophthalmol Vis Sci* 1997;38(5):817-23.

[158] Qi JH, Ebrahem Q, Moore N, Murphy G, Claesson-Welsh L, Bond M, et al. A novel function for tissue inhibitor of metalloproteinases-3 (TIMP3): inhibition of angiogenesis by blockage of VEGF binding to VEGF receptor-2. *Nat Med* 2003;9(4):407-15.

[159] Cruz-Munoz W, Kim I, Khokha R. TIMP-3 deficiency in the host, but not in the tumor, enhances tumor growth and angiogenesis. *Oncogene* 2006;25(4):650-5.

[160] Cha HS, Ahn KS, Jeon CH, Kim J, Song YW, Koh EM. Influence of hypoxia on the expression of matrix metalloproteinase-1, -3 and tissue inhibitor of metalloproteinase-1 in rheumatoid synovial fibroblasts. *Clin Exp Rheumatol* 2003;21(5):593-8.

[161] Ben-Yosef Y, Lahat N, Shapiro S, Bitterman H, Miller A. Regulation of endothelial matrix metalloproteinase-2 by hypoxia/reoxygenation. *Circ Res* 2002;90(7):784-91.

[162] Ottino P, Finley J, Rojo E, Ottlecz A, Lambrou GN, Bazan HE, et al. Hypoxia activates matrix metalloproteinase expression and the VEGF system in monkey choroid-retinal endothelial cells: Involvement of cytosolic phospholipase A2 activity. *Mol Vis* 2004;10:341-50.

[163] Canning MT, Postovit LM, Clarke SH, Graham CH. Oxygen-mediated regulation of gelatinase and tissue inhibitor of metalloproteinases-1 expression by invasive cells. *Exp Cell Res* 2001;267(1):88-94.

[164] Ridgway PF, Ziprin P, Alkhamesi N, Paraskeva PA, Peck DH, Darzi AW. Hypoxia augments gelatinase activity in a variety of adenocarcinomas in vitro. *J Surg Res* 2005;124(2):180-6.

[165] Kondo S, Kubota S, Shimo T, Nishida T, Yosimichi G, Eguchi T, et al. Connective tissue growth factor increased by hypoxia may initiate angiogenesis in collaboration with matrix metalloproteinases. *Carcinogenesis* 2002;23(5):769-76.

[166] Munoz-Najar UM, Neurath KM, Vumbaca F, Claffey KP. Hypoxia stimulates breast carcinoma cell invasion through MT1-MMP and MMP-2 activation. *Oncogene* 2006;25(16):2379-92.

[167] Fahling M, Perlewitz A, Doller A, Thiele BJ. Regulation of collagen prolyl 4-hydroxylase and matrix metalloproteinases in fibrosarcoma cells by hypoxia. *Comp Biochem Physiol C Toxicol Pharmacol* 2004;139(1-3):119-26.

[168] Chen PS, Zhai WR, Zhou XM, Zhang JS, Zhang YE, Ling YQ, et al. Effects of hypoxia, hyperoxia on the regulation of expression and activity of matrix metalloproteinase-2 in hepatic stellate cells. *World J Gastroenterol* 2001;7(5):647-51.

[169] Yamaguchi A, Tojyo I, Yoshida H, Fujita S. Role of hypoxia and interleukin-1beta in gene expressions of matrix metalloproteinases in temporomandibular joint disc cells. *Arch Oral Biol* 2005;50(1):81-7.

[170] Koong AC, Denko NC, Hudson KM, Schindler C, Swiersz L, Koch C, et al. Candidate genes for the hypoxic tumor phenotype. *Cancer Res* 2000;60(4):883-7.

[171] Koochekpour S, Jeffers M, Wang PH, Gong C, Taylor GA, Roessler LM, et al. The von Hippel-Lindau tumor suppressor gene inhibits hepatocyte growth factor/scatter factor-

induced invasion and branching morphogenesis in renal carcinoma cells. *Mol Cell Biol* 1999;19(9):5902-12.

[172] Petrella BL, Lohi J, Brinckerhoff CE. Identification of membrane type-1 matrix metalloproteinase as a target of hypoxia-inducible factor-2 alpha in von Hippel-Lindau renal cell carcinoma. *Oncogene* 2005;24(6):1043-52.

[173] Annabi B, Lee YT, Turcotte S, Naud E, Desrosiers RR, Champagne M, et al. Hypoxia promotes murine bone-marrow-derived stromal cell migration and tube formation. *Stem Cells* 2003;21(3):337-47.

[174] Feldmann M, Maini RN. Discovery of TNF-alpha as a therapeutic target in rheumatoid arthritis: preclinical and clinical studies. *Joint Bone Spine* 2002;69(1):12-8.

[175] Bainbridge J, Sivakumar B, Paleolog E. Angiogenesis as a therapeutic target in arthritis: lessons from oncology. *Curr Pharm Des* 2006;12(21):2631-44.

[176] Taylor PC, Paleolog EM. Is the vasculature a potential therapeutic target in arthritis? *Current Rheumatology Reviews* 2006;2(2):151-158.

[177] Miotla J, Maciewicz R, Kendrew J, Feldmann M, Paleolog E. Treatment with soluble VEGF receptor reduces disease severity in murine collagen-induced arthritis. *Lab Invest* 2000;80(8):1195-205.

[178] Afuwape AO, Feldmann M, Paleolog EM. Adenoviral delivery of soluble VEGF receptor 1 (sFlt-1) abrogates disease activity in murine collagen-induced arthritis. *Gene Ther* 2003;10(23):1950-60.

[179] de Bandt M, Grossin M, Weber AJ, Chopin M, Elbim C, Pla M, et al. Suppression of arthritis and protection from bone destruction by treatment with TNP-470/AGM-1470 in a transgenic mouse model of rheumatoid arthritis. *Arthritis Rheum* 2000;43(9):2056-63.

[180] Sone H, Kawakami Y, Sakauchi M, Nakamura Y, Takahashi A, Shimano H, et al. Neutralization of vascular endothelial growth factor prevents collagen-induced arthritis and ameliorates established disease in mice. *Biochem. Biophys. Res. Commun.* 2001;281(2):562-8.

[181] Lu J, Kasama T, Kobayashi K, Yoda Y, Shiozawa F, Hanyuda M, et al. Vascular endothelial growth factor expression and regulation of murine collagen-induced arthritis. *J. Immunol.* 2000;164(11):5922-7.

[182] Luttun A, Tjwa M, Moons L, Wu Y, Angelillo-Scherrer A, Liao F, et al. Revascularization of ischemic tissues by PlGF treatment, and inhibition of tumor angiogenesis, arthritis and atherosclerosis by anti-Flt1. *Nat Med* 2002;8(8):831-40.

[183] de Bandt M, Ben Mahdi MH, Ollivier V, Grossin M, Dupuis M, Gaudry M, et al. Blockade of vascular endothelial growth factor receptor I (VEGF-RI), but not VEGF-RII, suppresses joint destruction in the K/BxN model of rheumatoid arthritis. *J Immunol* 2003;171(9):4853-9.

[184] Grosios K, Wood J, Esser R, Raychaudhuri A, Dawson J. Angiogenesis inhibition by the novel VEGF receptor tyrosine kinase inhibitor, PTK787/ZK222584, causes significant anti-arthritic effects in models of rheumatoid arthritis. *Inflamm Res* 2004;53(4):133-42.

[185] Autiero M, Luttun A, Tjwa M, Carmeliet P. Placental growth factor and its receptor, vascular endothelial growth factor receptor-1: novel targets for stimulation of ischemic

tissue revascularization and inhibition of angiogenic and inflammatory disorders. *J Thromb Haemost* 2003;1(7):1356-70.

[186] Tjwa M, Luttun A, Autiero M, Carmeliet P. VEGF and PlGF: two pleiotropic growth factors with distinct roles in development and homeostasis. *Cell Tissue Res* 2003;314(1):5-14.

[187] Autiero M, Waltenberger J, Communi D, Kranz A, Moons L, Lambrechts D, et al. Role of PlGF in the intra- and intermolecular cross talk between the VEGF receptors Flt1 and Flk1. *Nat Med* 2003;9(7):936-43.

[188] Carmeliet P, Ferreira V, Breier G, Pollefeyt S, Kieckens L, Gertsenstein M, et al. Abnormal blood vessel development and lethality in embryos lacking a single VEGF allele. *Nature* 1996;380(6573):435-9.

[189] Carmeliet P, Moons L, Luttun A, Vincenti V, Compernolle V, De Mol M, et al. Synergism between vascular endothelial growth factor and placental growth factor contributes to angiogenesis and plasma extravasation in pathological conditions. *Nat Med* 2001;7(5):575-83.

[190] Maes C, Coenegrachts L, Stockmans I, Daci E, Luttun A, Petryk A, et al. Placental growth factor mediates mesenchymal cell development, cartilage turnover, and bone remodeling during fracture repair. *J Clin Invest* 2006;116(5):1230-42.

[191] Bottomley MJ, Webb NJ, Watson CJ, Holt L, Bukhari M, Denton J, et al. Placenta growth factor (PlGF) induces vascular endothelial growth factor (VEGF) secretion from mononuclear cells and is co-expressed with VEGF in synovial fluid. *Clin Exp Immunol* 2000;119(1):182-8.

[192] Oosthuyse B, Moons L, Storkebaum E, Beck H, Nuyens D, Brusselmans K, et al. Deletion of the hypoxia-response element in the vascular endothelial growth factor promoter causes motor neuron degeneration. *Nat Genet* 2001;28(2):131-8.

[193] Roskoski R, Jr. The ErbB/HER receptor protein-tyrosine kinases and cancer. *Biochem Biophys Res Commun* 2004;319(1):1-11.

[194] Woodburn JR. The epidermal growth factor receptor and its inhibition in cancer therapy. *Pharmacol Ther* 1999;82(2-3):241-50.

[195] Sweeney C, Carraway KL, 3rd. Negative regulation of ErbB family receptor tyrosine kinases. *Br J Cancer* 2004;90(2):289-93.

[196] Kusada J, Otsuka T, Matsui N, Hirano T, Asai K, Kato T. Immuno-reactive human epidermal growth factor (h-EGF) in rheumatoid synovial fluids. *Nippon Seikeigeka Gakkai Zasshi* 1993;67(9):859-65.

[197] Satoh K, Kikuchi S, Sekimata M, Kabuyama Y, Homma MK, Homma Y. Involvement of ErbB-2 in rheumatoid synovial cell growth. *Arthritis Rheum* 2001;44(2):260-5.

[198] Hallbeck AL, Walz TM, Briheim K, Wasteson A. TGF-alpha and ErbB2 production in synovial joint tissue: increased expression in arthritic joints. *Scand J Rheumatol* 2005;34(3):204-11.

[199] Kim JM, Ho SH, Park EJ, Hahn W, Cho H, Jeong JG, et al. Angiostatin gene transfer as an effective treatment strategy in murine collagen-induced arthritis. *Arthritis Rheum* 2002;46(3):793-801.

[200] Takahashi H, Kato K, Miyake K, Hirai Y, Yoshino S, Shimada T. Adeno-associated virus vector-mediated anti-angiogenic gene therapy for collagen-induced arthritis in mice. *Clin Exp Rheumatol* 2005;23(4):455-61.

[201] Kato K, Miyake K, Igarashi T, Yoshino S, Shimada T. Human immunodeficiency virus vector-mediated intra-articular expression of angiostatin inhibits progression of collagen-induced arthritis in mice. *Rheumatol Int* 2005;25(7):522-9.

[202] Cao Y. Endogenous angiogenesis inhibitors and their therapeutic implications. *Int. J. Biochem. Cell. Biol.* 2001;33(4):357-69.

[203] Cao Y, Chen A, An SS, Ji RW, Davidson D, Llinas M. Kringle 5 of plasminogen is a novel inhibitor of endothelial cell growth. *J. Biol. Chem.* 1997;272(36):22924-8.

[204] Sumariwalla P, Cao Y, Wu H, Feldmann M, Paleolog E. The angiogenesis inhibitor protease-activated kringles 1-5 reduces the severity of murine collagen-induced arthritis. *Arthritis Res Ther* 2003;5:R32-R39.

[205] Matsuno H, Yudoh K, Uzuki M, Nakazawa F, Sawai T, Yamaguchi N, et al. Treatment with the angiogenesis inhibitor endostatin: a novel therapy in rheumatoid arthritis. *J Rheumatol* 2002;29(5):890-5.

[206] Kurosaka D, Yoshida K, Yasuda J, Yokoyama T, Kingetsu I, Yamaguchi N, et al. Inhibition of arthritis by systemic administration of endostatin in passive murine collagen induced arthritis. *Ann Rheum Dis* 2003;62(7):677-9.

[207] Yin G, Liu W, An P, Li P, Ding I, Planelles V, et al. Endostatin gene transfer inhibits joint angiogenesis and pannus formation in inflammatory arthritis. *Mol Ther* 2002;5(5 Pt 1):547-54.

[208] Nagashima M, Asano G, Yoshino S. Imbalance in production between vascular endothelial growth factor and endostatin in patients with rheumatoid arthritis. *J Rheumatol* 2000;27(10):2339-42.

[209] Peacock DJ, Banquerigo ML, Brahn E. Angiogenesis inhibition suppresses collagen arthritis. *J Exp Med* 1992;175(4):1135-8.

[210] Oliver SJ, Banquerigo ML, Brahn E. Suppression of collagen-induced arthritis using an angiogenesis inhibitor, AGM-1470, and a microtubule stabilizer, taxol. *Cell Immunol* 1994;157(1):291-9.

[211] Oliver SJ, Cheng TP, Banquerigo ML, Brahn E. Suppression of collagen-induced arthritis by an angiogenesis inhibitor, AGM-1470, in combination with cyclosporin: reduction of vascular endothelial growth factor (VEGF). *Cell Immunol* 1995;166(2):196-206.

[212] Peacock DJ, Banquerigo ML, Brahn E. A novel angiogenesis inhibitor suppresses rat adjuvant arthritis. *Cell Immunol* 1995;160(2):178-84.

[213] Nagashima M, Tanaka H, Takahashi H, Tachihara A, Tanaka K, Ishiwata T, et al. Study of the mechanism involved in angiogenesis and synovial cell proliferation in human synovial tissues of patients with rheumatoid arthritis using SCID mice. *Lab Invest* 2002;82(8):981-8.

[214] Bernier SG, Lazarus DD, Clark E, Doyle B, Labenski MT, Thompson CD, et al. A methionine aminopeptidase-2 inhibitor, PPI-2458, for the treatment of rheumatoid arthritis. *Proc Natl Acad Sci U S A* 2004;101(29):10768-73.

[215] Fotsis T, Zhang Y, Pepper MS, Adlercreutz H, Montesano R, Nawroth PP, et al. The endogenous oestrogen metabolite 2-methoxyoestradiol inhibits angiogenesis and suppresses tumour growth. *Nature* 1994;368(6468):237-9.

[216] Leese MP, Hejaz HA, Mahon MF, Newman SP, Purohit A, Reed MJ, et al. A-ring-substituted estrogen-3-O-sulfamates: potent multitargeted anticancer agents. *J Med Chem* 2005;48(16):5243-56.

[217] Ireson CR, Chander SK, Purohit A, Perera S, Newman SP, Parish D, et al. Pharmacokinetics and efficacy of 2-methoxyoestradiol and 2-methoxyoestradiol-bis-sulphamate in vivo in rodents. *Br J Cancer* 2004;90(4):932-7.

[218] Newman SP, Leese MP, Purohit A, James DR, Rennie CE, Potter BV, et al. Inhibition of in vitro angiogenesis by 2-methoxy- and 2-ethyl-estrogen sulfamates. *Int J Cancer* 2004;109(4):533-40.

[219] Ho YT, Newman SP, Purohit A, Leese MP, Potter BV, Reed MJ. The effects of 2-methoxy oestrogens and their sulphamoylated derivatives in conjunction with TNF-alpha on endothelial and fibroblast cell growth, morphology and apoptosis. *J Steroid Biochem Mol Biol* 2003;86(2):189-96.

[220] Holmdahl R, Jansson L, Meyerson B, Klareskog L. Oestrogen induced suppression of collagen arthritis: I. Long term oestradiol treatment of DBA/1 mice reduces severity and incidence of arthritis and decreases the anti type II collagen immune response. *Clin Exp Immunol* 1987;70(2):372-8.

[221] Josefsson E, Tarkowski A. Suppression of type II collagen-induced arthritis by the endogenous estrogen metabolite 2-methoxyestradiol. *Arthritis Rheum* 1997;40(1):154-63.

In: New Research on Angiogenesis Inhibitors
Editor: Elmer T. Skinard, pp. 55-76
ISBN: 1-60021-473-8

Chapter III

ANTIANGIOGENIC PROPERTY OF METRONOMIC CHEMOTHERAPY

Katsumasa Kuroi[1,*] ***and Masakazu Toi***[2]

[1]Division of Surgery and Breast Oncology, Nyuwakai Oikawa Hospital, Fukuoka 810-0014, Japan;

[2]Division of Clinical Trial and Research, Tokyo Metropolitan Cancer and Infectious Disease Center, Komagome Hospital, Tokyo 113-8677, Japan.

ABSTRACT

Standard chemotherapy regimens are usually designed to kill as many tumor cells as possible by treating patients with the maximum tolerated dose (MTD) of cytotoxic agents. These agents are administered in a pulsed manner with breaks between cycles to allow recovery of normal tissues, thus limiting the toxicity of these agents. This strategy has led to a profound increase in the survival of children and adults with certain types of cancer. However, for the majority of solid tumors, particularly those at an advanced stage or associated with metastatic disease, neither complete tumor eradication nor sustained regression is often achieved. Moreover, MTD-based regimens are often associated with unpleasant or serious adverse effects that compromise the quality of life of the patient. The MTD concept targets the tumor cells themselves, and does not take into account the importance of the organization of a tumor, which consists of tumor cells and stromal tissue including endothelial cells (ECs).

In contrast, recent evidence suggests that most chemotherapeutic agents are capable of inhibiting tumor angiogenesis by targeting tumor ECs, and that metronomic, frequent or continuous administration of the same agent at low doses (a tenth to a third of the

* Correspondence concerning this article should be addressed to:
Dr. Katsumasa Kuroi Division of Surgery and Breast Oncology, Nyuwakai Oikawa Hospital, 2-21-16 Hirao, Chuo-ku, Fukuoka 810-0014, Japan TEL: +81-92-522-5411; FAX: +81-92-522-6244.
Dr. Masakazu Toi Division of Clinical Trial and Research, Tokyo Metropolitan Cancer and Infectious Disease Center, Komagome Hospital, 3-18-22 Honkomagome, Bunkyo-ku, Tokyo 113-8677, Japan. TEL: +81-3-3823-2101; FAX: +81-3-3824-1552.

maximum tolerated dose) might optimize its antiangiogenic properties. In fact, the results of preclinical and clinical studies have been encouraging, especially when drugs are used in combination with new molecularly targeted agents such as specific angiogenesis inhibitors. By targeting normal, terminally differentiated and genetically stable ECs, the risk of acquired drug resistance might be reduced. Through the use of metronomic chemotherapy, it may be possible to re-evaluate not only the re-treatment of refractory cancers with agents that have previously failed, but also agents that have traditionally been considered inactive or ineffective against a particular type of cancer. Moreover, as some of the most useful agents tested in both *in vitro* and *in vivo* models include vinca alkaloids, taxanes, and antimetabolites, the principle of this concept does not necessarily require the development of so-called "magic bullets".

At present, however, it remains difficult to determine the optimal doses and schedules for metronomic chemotherapy, as a rapid objective response is absent, and identification of MTD using conventional toxicity criteria is relatively straightforward. Also, it is still unclear whether metronomic chemotherapy actually exerts an antiangiogenic effect that contributes to putative antitumor efficacy, and there is currently no standard for monitoring and quantifying the angiogenic response during therapy. Nevertheless, from a clinical view point, metronomic chemotherapy appears promising, not only because it reduces toxicity and the need for growth-factor support or anti-emetic drugs, but also because it may improve antitumor effects. The actual significance of this concept should be confirmed by much larger, prospective, and randomized controlled trials that compare metronomic dosing chemotherapy schedules with conventional ones. In addition, novel functional or angiogenesis-specific markers and new high-resolution techniques for blood flow imaging and analysis are essential for monitoring of angiogenesis.

INTRODUCTION

The concept of standard chemotherapy is to administer an agent at a dose close to the maximum that a patient can tolerate before the onset of unacceptable toxicity. This is based on the precept that tumor cells represent the prime target of chemotherapy and that eradication can be achieved only through repetitive administration of a conventionally derived maximum tolerated dose (MTD). On the other hand, recent evidence suggests that most chemotherapeutic agents are capable of inhibiting tumor angiogenesis by targeting tumor endothelial cells (ECs), and that metronomic, frequent or continuous administration of the same agent at low doses might optimize its antiangiogenic properties [1-4]. These findings have broken out of the paradigm shift of anticancer chemotherapy, and now tumor ECs have become an important target of chemotherapy. In this article, we summarize the current status of knowledge about the antiangiogenic properties of metronomic, continuous low-dose chemotherapy.

Current Concepts of the Tumor Growth Model and Standard Chemotherapy

MTD-based chemotherapy is based largely on the "much is better" concept (Table 1). This was originally derived from kinetic modeling of curative chemotherapy response of murine leukemia L1210 cells *in vitro*, which are a cell line showing rapid and almost exponential growth fraction closed to 100% as a consequence of frequent cell cycling [5,6]. In this model, the cytotoxic effects of chemotherapy follow a log-kill kinetics model whereby a given dose always kills a certain fraction, rather than a certain number, of exponentially growing tumor cells. This model is attractive because of its simplicity, and thus modern chemotherapy regimens are usually designed to kill as many tumor cells as possible by treating patients with the highest doses possible without causing life-threatening levels of toxicity, *i.e.* the MTD. In general, host toxicity is often only marginally less than anti-tumor efficacy, creating a narrow therapeutic index, and the agents are usually administered in a pulsed manner with breaks (usually of 3 to 4 weeks) between cycles to allow recovery of normal dividing cells of rapidly regenerating tissues such as bone marrow and gut mucosa, at the sacrifice of hair-follicle cells. These regimes are not only associated with unpleasant or serious adverse effects that compromise the quality of life for the patient who is undergoing therapy, but also provide a chance for tumor regrowth during the resting period. The apex of this concept is high-dose chemotherapy with the use of autologous hematopoietic stem cell support.

MTD-based chemotherapy has led to a profound increase in the survival of children and adults with certain types of cancer, whereas responses for the majority of solid tumors, particularly when they are advanced or associated with metastatic disease, are short lived, and neither complete tumor eradication nor sustained regression is often achieved. In this respect, it is important to note that the MTD concept was developed using synchronized, log-phase non-mutagenic cells and non-cell-cycle-specific agents for which chemotherapy was optimized to affect all cells in culture. Therefore, this concept is unlikely to reflect the nature of most solid tumors, which consist of heterogeneous cell populations, where only cells at a specific stage of the cell cycle are likely to be affected by chemotherapy. In addition, genetic and epigenetic variations within a heterogeneous population of tumor cells might result in diversity of drug sensitivity, thus eventually conferring refractoriness to conventional chemotherapy. For these reasons, most human solid tumors do not exhibit purely exponential growth, but rather appears to show non-exponential Gompertzian kinetics [7]. In this model, the ratio of proliferating cells to total cells decreases with increasing tumor size, the growth fraction declines exponentially over time, and a plateau phase can be observed in some malignancies. As a result, the tumor growth curve has three distinct regions; a slow initial phase, a middle phase characterized by rapid growth, and a slow plateau phase. Another important consideration in this model is the impact of residual tumor cells on tumor growth after chemotherapy. In volume-reduced Gompertzian cancer models, regrowth of tumor cells between cycles of chemotherapy is more rapid than in exponential models, and is fastest at small tumor volumes.

Table 1. Comparison of several concepts of chemotherapy

	Conventional chemotherapy	Dose-dense chemotherapy	High-time chemotherapy	Metronomic chemotherapy
Primary aim	Total tumor cell kill	Total tumor cell kill	Maximum tumor cell kill	Dormancy status
Dosage	MTD-based	MTD-based	MTD or DLT of longer exposure, lower-dose schedule	1/3 to 1/10 of MTD
Dose intensity	High	High	Low	Low
Interval	Intermittent (usually 3 weeks)	Intermittent (usually 1-2 weeks)	Daily / continuous	Continuous (daily / week)
Duration	Short term	Short term	Long term	Long term
Projected target	Tumor cells	Tumor cells	Not defined (Endothelial cells for low dose)	Endothelial cells
Endpoint in clinical study	Established	Established	Feasibility of long term administration	Not established
Levels of toxicity	High	High	Low	Low
Quality of life	Impair	Impair	Maintain	Maintain
Resistance	Frequent	Frequent	Not defined	Rare
Need for growth factor support	Sometimes	Frequently / mandatory	Available	No
Combination with anti-angiogenic or targeted drugs	Available	Available	Available	Available

Abbreviations: DLT, dose-limiting toxicity; MTD, maximum tolerated dose.

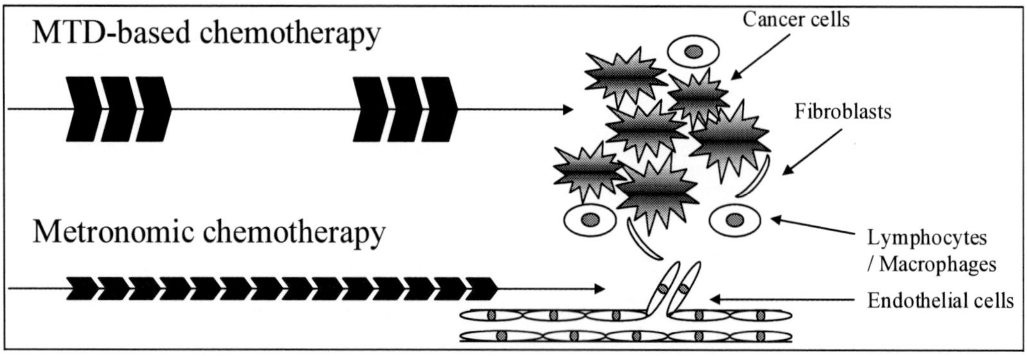

Fig. 1. Target of standard chemotherapy and metronomic chemotherapy. Tumor tissue consists of tumor cells and stroma including endothelial cells. The standard chemotherapy mainly targets tumor cells, while metronomic chemotherapy targets tumor endothelial cells. The size of the bullets and interspaces refer to the doses and interval of chemotherapy, respectively.

Based on this concept, Norton [8] has proposed that administration of an appropriate and effective dose of chemotherapy more frequently would limit tumor regrowth between cycles and minimize the possibility of acquired chemoresistance with a higher probability of logarithmic tumor-cell destruction. This approach is referred to as dose-dense chemotherapy, and is usually designed so that at least the same amount or, more commonly, even a greater total amount of drug can be administered over time to reach the highest possible dose, resulting in an increase of dose intensity (DI). In agreement with this hypothesis, it has been reported that weekly dosing of paclitaxel involving delivery of lower individual, but more frequent doses of paclitaxel was effective for the treatment of advanced breast cancer [9]. Another example is the Intergroup Trial C9741 / Cancer and Leukemia Group B (CALGB) Trial 9741. This trial was designed to test two concepts - dose density and sequential therapy - based on experimental data and mathematical reasoning, and demonstrated that dose-dense chemotherapy improved survival relative to standard chemotherapy in an adjuvant setting for patients with node-positive breast cancer [10].

In either case, however, the MTD concept targets tumor cells themselves, and does not take into account the importance of individual tumor components including tumor cells and stromal tissue containing ECs [11-13], despite the fact that angiogenesis is a hallmark characteistics of cancer, and that tumor growth and metastasis are dependent on continued angiogenesis (Fig. 1).

ALTERNATIVE CONCEPT OF CHEMOTHERAPY

In chemotherapy, from a basic pharmacological viewpoint, the principles of response lag (threshold), response slope and plateau are evident, and the influence of exposure duration is unmistakable [14]. Indeed, the often-cited formula for cell killing, e^{-kct}, has both a concentration (c) and a time (t) component in the exponent, and thus one can hypothesize that once a killing effect has been achieved with a maximal dose, then no further efficacy can been expected, the only remaining variable is time. Recently, the importance of scheduling has been rediscovered as a new "old" concept of chemotherapy, termed high-time chemotherapy [2]. This chemotherapy approach seeks the maximum drug exposure time for a given drug concentration. For example, when an effective dose of a drug can be maintained for 14 days with little and reversible toxicity, we can either increase the daily dose during this period, or extend the administration period of the same dose to 21 or more days until the longest possible time that results in dose-limiting toxicity (DLT). Although this concept is useful for long-term therapy, careful assessment of the pharmacokinetics and pharmacodynamics of the drug involved would be required [4].

On the other hand, since recent evidence has suggested that dividing ECs are present in the growing blood vessels formed in tumors [15], there is now a shift in thinking towards the view that either continuous infusion or frequent, even daily, administration of chemotherapeutic agents at doses significantly below the MTD with no prolonged drug-free breaks would be more effective, not only in terms of reducing certain toxicities but perhaps even improving the antitumor effect as well, by shifting the target of the chemotherapy from the tumor cells to the vascular ECs [2-4]. This concept was first referred to by the term "metronomic chemotherapy" by Hanahan *et al.* [3], while the term "antiangiogenic dosing" or "high-time for low dose chemotherapy" have been advocated by others [2,16,17]. In either case, targeting normal, terminally differentiated and genetically stable ECs may have a theoretical advantage in avoiding, or at least delaying the onset of acquired resistance which, in part, occurs because of the massive and diverse genetic instabilities present in tumor cells. In other words, drug resistance could be reversed simply by shifting the focus of the treatment from the tumor cells to still-sensitive tumor ECs. This is supported by the fact that the maximum level of myelosuppression observed in cancer patients remains unchanged after multiple cycles of chemotherapy, even when the tumors in these patients have become drug-resistant [18]. If normal bone marrow progenitor cells acquire resistance to chemotherapy in the same way that genetically unstable, highly mutable tumor cells do, then myelosuppression would gradually decline and disappear.

In addition to direct effect on ECs, several studies have indicated that an endogenous inhibitor of angiogenesis, thrombospondin 1 (TSP1), is a potent mediator of the effect of metronomically administered cyclophosphamide (CPA) [19-21]. TSP1 blocks the proliferation of, and induces apoptosis in ECs by binding to CD36 receptors, which are expressed by ECs [22]. TSP1 can also bind and sequester vascular endothelial growth factor (VEGF), and thus block its proangiogenic activity. Moreover, a proapoptotic siganal elicited by TSP1 generates CD95L, a ligand for the CD95 death receptor [23]. In addition to CPA, several other chemotherapeutic agents such as cisplatin, docetaxel, and doxorubicin could induce endothelial CD95 in vivo and in vitro at low doses that do not kill ECs [24,25]. In contrast, as metronomically administered CPA retains its efficacy in mice that are unable to produce the other endogenous angiogenesis inhibitors endostatin and tumstatin [19], these molecules might not be involved in the antiangiogenic effect of this type of chemotherapy. Thus, metronomic chemotherapy might not necessarily act directly on ECs, but might exert its effect by inducing the EC-specific inhibitor TSP1 [26,27].

Interestingly, the metronomic approach does not appear to be confined to chemotherapy. For example, tumor response to radiotherapy is regulated not only by tumor cell phenotype, but also by EC apoptosis. In a study by Garcia-Barros *et al.* [28], antitumor effects of irradiation were mediated through a primary event involving damage or destruction of the tumor neovasculature, followed by the death of tumor cells surrounding the affected vessels, indicating that microvascular sensitivity regulates tumor cell response to radiation within the clinically relevant dose range. In fact, radiation is sometimes administered at a lower than normal dose, known as hyperfractionated radiation. Moreover, in a study by Slaton *et al.* [29], daily administration of smaller doses of interferon (IFN)-alpha was significantly more effective than larger doses once a week in a mouse cancer xenograft model. IFN-alpha is also effective for treatment of pediatric patients with hemangiomas or giant-cell tumors when administered in small daily doses over prolonged periods of time [30,31]. Similarly, metronomic photodynamic therapy has been shown to be useful in a malignant brain tumor model [32,33].

On the other hand, more compressed or accelerated schedules of metronomic chemotherapy might be viewed as a variation of dose-dense chemotherapy. However, metronomic chemotherapy treats cancer as a chronic disease in which the tumor burden is maintained at the lowest achievable level over a long period, whereas the latter aims for complete tumor cell killing based on the MTD-concept, as mentioned above. Thus, metronomic chemotherapy is a different approach from dose-dense chemotherapy for the management of cancer, and the concept of metronomic chemotherapy is supported by several mathematical modeling studies [34,35].

Reasons for Ineffective Antiangiogenic Activity of MTD-Based Chemotherapy

As tumor ECs are not quiescent, and moderately high fractions of proliferating ECs can be detected during tumor angiogenesis [15], dividing ECs should be vulnerable to the cytotoxicity of chemotherapeutic agents. However, there might be some reasons why MTD-based chemotherapy does not effectively target ECs. First, MTD-based chemotherapy usually necessitates breaks in drug administration to allow bone marrow recovery, and thus damage to ECs could be largely repaired during such rest periods. In fact, in a study by Browder *et al.* [16], CPA was shown to cause apoptosis of ECs when administered at the MTD in immune-competent syngeneic mice. In this study, the first cells in the tumor to undergo apoptosis were the ECs, although this antiangiogenic effect was of no significant therapeutic benefit.

In association with this, it is important to note that chemotherapy could mobilize circulating endothelial progenitors cells (CEPs) from the bone marrow into the circulation, and these mobilized CEPs would be recruited into the tumor vasculature [34,36,37]. In fact, a recent study using two human lymphoma xenograft models has demonstrated that low-dose metronomic dosing of CPA caused a sustained reduction of viable CEPs, whereas MTD-based administration caused a marked drop followed by a rapid rebound during the break period [38]. Thus, the prolonged interval between chemotherapy cycles would provide an opportunity for new ECs to be recruited, either from existing vasculature or from CEPs, allowing angiogenesis to proceed [38-40]. Alternatively, the ECs might be protected from damage or cell death induced by chemotherapeutic agents by high local concentration of EC survival factors such as VEGF, basic fibroblast growth factor, and thymidine phosphorylase [41-43].

Thus, little therapeutic benefit might be derived from this transient antiangiogenic effect, and in fact, conversly, endothelial recovery during the treatment-free period might support the regrowth of tumor cells [16]. This concern has been shown to be justified in a pivotal trial conducted in patients with metastatic colorectal cancer; survival was improved in patients receiving irinotecan, fluorouracil, and lecovorin with the anti-VEGF agent bevacizumab, in comparison with patients who did not receive bevacizumab [44]. Adding bevacizumab might alter the tumor growth pattern, improve intratumoral chemotherapy delivery, and result in improved rates of minimal residual disease and possibly disease eradication. This finding is consistent with the Norton-Simon hypothesis that addition of any effective anticancer agent to chemotherapy could improve results by limiting tumor growth between chemotherapy cycles, as tumor regrowth requires the development and maintenance of tumor vasculature and the recruitment of a blood supply [45].

Preclinical Evidence of Metronomic Chemotherapy

The strategy of metronomic scheduling of chemotherapy to exploit its capacity to inhibit angiogenesis was first supported by experimental evidence that the use of a lower dose of CPA (approximately one-third the MTD for mice) with more frequent weekly administrations reduced the risk of acquired drug resistance in mice bearing Lewis lung carcinoma and L1210 leukemia cells [16]. This strategy facilitated tumor eradication by denying any opportunity to repair the damage inflicted on the vasculature. It is also interesting that mouse tumors which had previously acquired resistance to CPA administered using a conventional MTD schedule responded dramatically over time to the same drug when a lower dose, with a more frequent schedule, was started, and addition of the angiogenesis inhibitor TNP-470 to the antiangiogenic schedule of CPA eradicated drug-resistant tumors. The same study also confirmed that metronomic dosing of a number of other cytotoxic agents including 5-fluorouracil had an antiangiogenic effect.

Similarly, Klement *et al.* [46] demonstrated that the mitosis-blocking cytotoxic agent vinblastine directly killed cultured ECs at doses considerably below those required to affect drug-sensitive neuroblastoma cells, and that intraperitoneal administration of low-dose vinblastine well below the MTD with a continuous schedule impaired tumor growth in mice. The same study also explored the use of the monoclonal antibody DC101 to disrupt the function of the VEGF type 2 receptor (VEGFR), which is highly expressed on activated ECs. Interestingly, the combination of the antiangiogenic vinblastine with DC101 resulted in complete and sustained regression of established neuroblastoma xenografts in mice, while DC101 alone resulted in significant but transient tumor regression. Histopathological assessment revealed that vinblastine alone and DC101 alone both increased the width of the apoptotic rim, suggesting that cells most distal to the tumor vasculature were primarily affected, although a large percentage of tumor cells still remained viable in the tumor center, and despite an evident increase in apoptosis, net tumor growth was observed. In contrast, in the combined therapy group, there was overwhelming loss of both cell viability and preexisting tumor architecture. Vascular damage was common to all the treatment groups, but the prevalence and severity of the damage were greatest in the combination treatment group. In another study, the same group also demonstrated that continuous low-dose (one-tenth of MTD) chemotherapy was highly effective against orthotopically grown multidrug-resistant human breast cancers in mice with severe combined immunodeficiency disease, but usually only when the drugs were used in combination with DC101 [47]. In this study, paclitaxel and vinblastine exhibited better therapeutic profiles than other drugs including cisplatin or doxorubicin, mainly because of their lack of cumulative toxicity such as weight loss.

Moreover, Man *et al.* [48] found that low-dose CPA administration in drinking water was a safe and convenient treatment with significant antitumor efficacy in xenotransplanted Swiss nude mice carrying tumors derived from human breast, colon and melanoma cell lines. This study clearly demonstrated that per os drug in human could be chronically and safely administered in a similar but more convenient basis in mice through drinking water, and similar to the previous studies, the effects of continuous administration of CPA was enhanced by DC101.

Collectively, these studies have established that a metronomic chemotherapy dosing schedule can be antiangiogenic, achieve a durable and potent tumor response, and also provide a stable and safe way to circumvent multidrug resistance, simply by shifting the focus of the treatment away from the drug-resistant tumor cell population to the drug-sensitive tumor ECs. At the same time, these studies have demonstrated the value of using angiogenesis inhibitors in combination with the main therapy. These results have since been confirmed by others with or without combination with targeted drugs such as thalidomide, PEX fragment of metalloprotease-2, platelet factor fragment PF-4, anti-endogrin antibody, TSP1 peptide, endostatin, anti-VEGFR, kinase inhibitors (imatinib, SU5416, SU11248), metalloprotease inhibitor (BB-94, BAY 12-9566), immunotherapy, or hyperthermia [24-26,49-64].

Among them, Hermans *et al.* [58] assessed a combination of metronomic administration of CPA with specific antitumor immunotherapy eliciting cytotoxic T lymphocytes (CTLs) reactive to recombinant epitopes expressed by the murine melanoma model B16.F10, and found that this combination strategy dramatically enhanced antitumor responses over either therapy alone. Both metronomic chemotherapy and MTD-based chemotherapy deleted the proliferating tumor-specific CTLs in blood, although the deletion showed slower kinetics with the metronomic schedule. This study demonstrated the potential benefits of combining non-immunosuppressive metronomic dosing of CPA with an immunotherapeutic approach to treat cancer. Moreover, in a study by Ma *et al.* [63], metronomic dosing of gemcitabine and cisplatin significantly attenuated the disruption of the homeostatic balance between tissue factor and the tissue factor pathway, suggesting that metronomic chemotherapy might reduce adverse clotting events associated with chemotherapy alone or in conjunction with antiangiogenic drug combination therapies. In association with this, it is interesting to note that, in a study by Emmenegger *et al.* [65], long-term daily low-dose CPA administration caused neither significant toxicity to bone marrow and gut mucosa normally affected by MTD-based chemotherapy of the same agent, nor had a negative impact on wound healing.

Moreover, a study of particular interest investigating the "chemo-switch" protocol, a MTD-based chemotherapy followed by metronomic chemotherapy as maintenance therapy, demonstrated that the inhibition of platelet-derived growth factor receptor signaling by the tyrosine kinase inhibitors imatinib and SU11248 enhanced the efficacy of metronomic chemotherapy in a transgenic mouse model of pancreatic neuroendocrine cancer [59]. Treatment with imatinib reduced the pericyte coverage of tumor blood vessels, rendering ECs more sensitive to chemotherapy. Moreover, using a combination of imatinib and SU11248 with metronomic chemotherapy elicited regression of established tumors and prolonged the survival of the mice. By combining an initial MTD-based chemotherapy with maintenance therapy consisting of metronomic chemotherapy and SU11248, remarkable efficacy was achieved in terms of objective response and survival. Importantly, this study highlights the significance of pericytes as a target of antiangiogenic therapy, and validates the chemo-switch regimen. In addition, in a study by Shaked *et al.* [64], repeated administration of bolus doses of CPA every 3 or 6 weeks, combined with a daily oral low-dose metronomic regimen, improved efficacy and significantly delayed the progression of transplanted PC-3 human prostate cancer xenografts, syngenic transplanted EMT-6 breast tumors, and "spontaneous" murine erythroleukemia. In this study, the bolus dose and low-dose metronomic chemotherapy caused a sustained reduction in the viability of CEPs. Thus, standard MTD-based "remission-induction" chemotherapy and long-term metronomic chemotherapy might not be mutually exclusive, and could be used in a beneficial and harmonious manner [27].

Clinical Precedents of Metronomic Chemotherapy

A retrospective view of medical science reveals that there are actually many clinical precedents for these preclinical findings. For example, some of the most successful treatment regimens in pediatric oncology resemble metronomic chemotherapy in that they involve daily administration of low doses of cytotoxic agents over prolonged periods of time as maintenance therapy. In fact, in the protocol for acute lymphoblastic leukemia, children are treated with low doses of oral methotrexate on a weekly basis and 6-mercaptopurine on a daily basis for up to 3 years [27,66]. In this regard, it is interesting to note that drugs used in this setting have been demonstrated to be antiangiogenic [67-69]. In addition, daily administration of CPA and weekly administration of vinca alkaloids such as vincristine or vinblastine has been shown to be effective for treatment of infants with neuroblastoma [70], and weekly chronic chemotherapy using vincristine and actinomycin, or vincristine, actinomycin and CPA, for nearly half a year has also been used successfully for the treatment of pediatric patients with rhabdomyosarcoma [71]. Moreover, the pediatric regimen (CPA, doxorubicin, vincristine and prednisone), which is used for treatment children with non-Hodgkin's lymphoma is far less intensive in the use of vincristine and prednisone than the regimen for adult, and is conceptually similar to metronomic chemotherapy [27].

Similarly, a variety of chemotherapeutics resembling an antiangiogenic dosing schedule has also been documented for adult solid and hematological tumors. For example, in a study comparing the classical CPA, methotrexate, and 5-fluorouracil (CMF) schedule with a modified 3-week intravenous CMF schedule for postmenopausal patients with advanced breast cancer, the classical regimen resulted in a significantly better response rate and survival, possibly because, in the classical CMF schedule, CPA is administered orally on days 1-14, followed by 2 weeks rest, which more closely resembles an antiangiogenic schedule than intravenous administration of CPA every 3 weeks. Similarly, oral administration of uracil and tegafur (UFT) and leucovorin daily for 28 days every 35 days in patients with metastatic colorectal cancer was reportedly well tolerated, without causing neutropenia and oral mucositis that complicates intravenous schedules of 5-fluorouracil plus leucovorin [72]. In this context, it is important to note that oral administration of UFT is commonly used to treat a wide variety of solid tumors in Japan [73,74], and the usefulness of oral UFT has also been demonstrated in an adjuvant setting in patients with colon cancer or breast cancer [74-76].

Moreover, intuitive physicians have found that significant proportions of patients with breast and ovarian cancer who had stopped responding to MTDs of a taxane given once every 3 weeks, subsequently responded to the same drug once it was switched to a weekly schedule at about one third of the MTD [77-79]. These weekly schedules using lower doses were found to minimize the toxicities associated with the once-every-3-week MTD regimen [78-82]. Similar results have been observed in patients refractory to a bolus MTD dose of agents such as etoposide and CPA when the same agent was administered chronically at a lower dose [83]. Furthermore, there are several other situations in which prolonged oral administration of relatively low doses of chemotherapeutics is already in use for palliative-like regimens that are less toxic and are sometimes used to treat elderly patients who are less able to cope with MTD-based chemotherapy. These agents include etoposide [84,85], an antiangiogenic topoisomerase II inhibitor, razoxane [86], an alkylating agent, temozolamide [87-89], a prodrug of a bifunctional alkylating agent, treosulfan [90], and an alkylating oxazaphosphorine, trofosfamide [91]. Among them, razoxane was initially developed as a potent antiangiogenic agent, and was later found to be a noncleavable inhibitor of topoisomerase II. Similarly, low-dose methotrexate plus vinblastine has been administered every 7-10 days for several months in a substantial subset of patients with advanced, inoperable aggressive fibromatosis [92]. Although it is has not yet been demonstrated conclusively whether these regimens described above actually have antiangiogenic activity that contributes to their antitumor efficacy, the primary importance of these studies appears to be that a metronomic-like schedule of chemotherapy is feasible and of potential clinical benefit.

Clinical Studies of Metronomic Chemotherapy

Several clinical studies including case reports, retrospective studies, feasibility studies and phase II studies have evaluated the possible benefits of metronomic chemotherapy regimens (Table 2). In most of these, CPA has been involved, and administered orally on a daily basis, sometimes for up to 2 years, with no break periods, and in some cases, oral low-dose methotrexate has also been given on two consecutive days on a weekly basis. For example, the phase II study conducted by Colleoni *et al.* [93] evaluated low-dose oral CPA (50 mg/day) administered on a daily basis and oral methotrexate (2.5 mg, bd) administered twice a week in 64 patients with progressive, advanced and refractory breast cancer, and achieved an overall response of 32% with no high-grade toxicity. In this study, serum VEGF levels decreased two months after the start of treatment, although there was no significant difference in the median reduction between responders and non-responders.

On the other hand, Globe *et al.* [94] performed a chart review to evaluate the efficacy and toxicity of continuous oral administration of a combination of CPA (50 mg/day) and dexamethasone (1 mg/day) in patients with prostate-specific antigen (PSA) progression despite hormone therapy and antiandrogen withdrawal. Dexamethasone, in addition to its other properties, has been shown to exert an antiangiogenic effect, possibly by through inhibiting the production of VEGF [95,96]. They reported that almost 70% of the patients showed a decrease in serum PSA levels of 50% or more, and that the duration of response was 8 months with well tolerated toxicities.

Similarly to preclinical studies of metronomic chemotherapy, several studies have combined metronomic chemotherapy with targeted drugs such as cyclooxygenase-2 inhibitors (celecoxib, rofecoxib), the peroxisome proliferator-activated receptor gamma agonist (pioglitazone), thalidomide [97-105], and more recently, designated angiogenesis inhibitors such as bevacizumab [27,106]. It is also interesting to note that metronomic chemotherapy was found to be cost effective in terms of outcome and resource utilization when low-dose CPA and methotrexate chemotherapy was compared with published phase II studies using oral or intravenous chemotherapy in a palliative setting for previously treated metastatic breast cancer [105].

Table 2. Reported studies of metronomic chemotherapy.

Reporter, year	Patient population, setting	Treatment	Outcome
Colleoni M, 2002 [93]	Metastatic breast cancer, phase II study (n=64)	Low-dose CPA (50 mg/day) PO, and low-dose MTX (2.5 mg bd, twice a weekly) PO	2 CR , 10 PR 10, 8 SD for at least 24 weeks, 9 grade 3 liver toxicity, 12 grade 2 leukopenia, decrease of serum VEGF level at 2 months
Glode LM, 2003 [94]	Advanced hormone refractory prostate cancer, retrospective study (n=34)	Low-dose CPA (50 mg/day) and low-dose dexamethasone (1 mg/day)	22 CR, 2 SD assessed by prostate specific antigen response, duration of treatment 9 months
Spieth K, 2003 [97]	Advanced melanoma, pilot study (n=12)	Low-dose oral treosulfan (500 mg/day) , and cyclooxygenase-2 inhibitor, rofecoxib (25 mg/day)	1 PR, 4 SD (12-36 weeks), median survival time 13 months, no grade 3, 4 toxicities
Vogt T, 2003 [98]	Advanced and pretreated but progressive, malignant vascular tumors, pilot study (n=6)	Oral pioglitazone (45 mg/day), rofecoxib (25 mg/ day), and after 14 days, oral trofosfamide (3 x 50 mg/day)	CR 2, 1 PR, SD 3, median progression-free survival 7.7 months, mild toxicities
Coras BC, 2004 [99]	Endemic Kaposi sarcoma, a case report (42 years old)	Same as the regime reported by Vogt	PR for 18 months, no significant toxic effects
Nicolini A, 2004 [100]	Metastatic hormone refractory prostate cancer, case series (n=8)	Low-dose oral CPA (100-150 mg/day, alternately)	2 PR, 3 SD, 3 PD, grade 2, 3 neutropenia in all patients
Reichle A, 2004 [101]	Previously treated metastatic melanoma or soft tissue sarcoma, phase II study (n=40)	Same as the regime reported by Vogt	4 CR, 2 PR, 5 SD for more than 6 months, no grade 3, 4 toxicities
Fassas AB, 2005 [102]	Refractory Waldenstrom's macroglobulinemia, a case report	Low-dose chemotherapy with continuous intravenous CPA and adriamycin, and rituximab, dexamethasone, bortezomib, thalidomide, rapamycin, interferon-gamma	Minimal lymphomatous involvement in bone marrow, stable monoclonal markers, normal platelet count, marked functional improvement
Herrlinger U, 2005 [103]	Recurrent glioblastoma, phase II (n=10)	Oral low-dose CPA (100 mg/day) and MTX (5mg twice weekly)	No CR, no PR, prematurely closed because progression-free survival rate at 6 months was 0%
Kieran MW, 2005 [104]	Recurrent or progressive pediatric cancer, feasibility study (n=20)	Thalidomide, celecoxib with alternating oral etoposide and CPA every 21 days for 6 months	3 PR, 40% of patients completed 6 months of treatment, correlation of baseline level of thrombobospondin-1 and prolonged response, no response pattern in VEGF, bFGF, endostatin
Bocci G, 2005 [105]	Metastatic breast cancer, comparative pharmacoeconomic evaluation	Low-dose CPA and MTX	Metronomic chemotherapy was significantly cost-effective

Abbreviations: bFGF, basic fibroblast growth factor; CPA, cyclophosphamide; CR, complete response; DLT, dose-limiting toxicity; PO, per os; SD, stable disease; PD, progressive disease; PR, partial response; MTX, methotrexate; VEGF, vascular endothelial growth factor.

Current Issues and Future Directions of Metronomic Chemotherapy

Since Polverini *et al.* [107] first reported the antiangiogenic effect of conventional chemotherapy as early as 15 years ago, virtually every class of chemotherapeutic agent has been shown to have antiangiogenic activity, as determined by *in vitro* and *in vivo* assays, using multiple potential criteria for antiangiogenicity, including differential cytotoxicity to ECs at doses that have no effect on tumor cells themselves, interference with EC function without EC death, interference with a specific portion of the angiogenic cascade, and *in vivo* evidence of antiangiogenic activity [1]. However, some agents have profound antiangiogenic effects at very low doses, others exhibit antiangiogenic activity only at near-maximum cytotoxic concentrations, whereas others have striking antiangiogenic activity in one xenograft model but none in others. Moreover, only a few agents have been demonstrated to produce substantial antiangiogenic activity *in vivo*, while several cytotoxic agents have been shown to affect ECs *in vitro* [4]. In contrast, metronomic dosing of some agents could stimulate VEGF-mediated angiogenesis. In a study by Albertsson *et al.* [108], metronomic cisplatin and fluorouracil treatment significantly stimulated angiogenesis in a dose-dependent, non-linear manner *in vivo*. The proangiogenic effect occurred frequently in the case of cisplatin and infrequently in the case of fluorouracil. Thus, there might be a finely tuned dose-effect situation, and no set criteria could actually define the antiangiogenic activity of cytotoxic agents.

From the viewpoint of clinical practice, there are a number of challenges to devising metronomic chemotherapy regimens in terms of the choice of agents, dosing schedule, and study design. For example, as rapid objective responses are absent and identification of MTD using conventional toxicity criteria is relatively straightforward for metronomic chemotherapy, selection of an optimum antiangiogenic dose that is nontoxic but still effective remains difficult. In addition, the critical pathways by which EC biology can be manipulated have not been identified *in vivo*, and levels of circulating angiogenesis regulators have not proven particularly effective as markers of antiangiogenic therapy [109-111]. Thus, there is still no standard for monitoring and quantifying the angiogenic response during therapy. Therefore, the approach of metronomic chemotherapy remains largely empirical. In fact, the dose for metronomic chemotherapy has been empirically determined to be 10 – 33% of conventional doses, in a similar way to conventional chemotherapy where the starting dose of a phase I study is empirically determined to be 1 to 0.1 of the LD_{50} obtained in a preclinical, animal-based study [112, 113]. In this respect, it is interesting to note that in a recent study by Shaked *et al.* [114], levels of CEPs or circulating ECs (CECs) correlated well with the degree of tumor angiogenesis and the response to VEGFR-targeted antiangiogenic therapy. In another study, the same group also found that each optimum biologic dose was strikingly correlated with the maximum reduction in viable CEPs [115]. Thus, monitoring of CEPs or CECs might provide a sensitive index [114,116-119], whereby the highest dose that can be administered given in a metronomic manner would be the dose exerting the greatest effect on ECs without clinical bone marrow perturbation [112,115]. Another promising possibility would be to determine the optimal dose level for a drug by detecting the circulating levels of TSP1.

In summary, metronomic chemotherapy appears to be promising, not only because it reduces toxicity and the need for growth factor support or anti-emetic drugs, but also show better anti-tumor effects. When MTD-based chemotherapy is indicated, however, it should not be changed just for the sake of increasing the antiangiogenic efficacy of any given agent until the actual significance of the metronomic concept has been confirmed by much larger, prospective, and randomized controlled trials comparing metronomic dosing schedules with conventional chemotherapy. It is also important to clarify which agents work best for metronomic chemotherapy, which combination or sequence with other modalities such as antiangiogenic agents, molecular-targeted agents or immunotherapy are more effective for the treatment of different tumor entities, or what mechanisms of resistance might develop over time. In theory, endocrine therapy might be an additional choice for advantageous combination with metronomic chemotherapy, as recent studies have indicated that 17 beta-estradiol modulates VEGF and VEGFR expression [120,121]. Moreover, novel functional or angiogenesis-specific markers and new high-resolution techniques for blood flow imaging and analysis are vital for monitoring antiangiogenic activity and response *in vivo*. Finally, careful monitoring of long-term or delayed adverse effects of metronomic chemotherapy will be necessary.

References

[1] Miller KD, Sweeney CJ, Sledge GW, Jr. Redefining the target: chemotherapeutics as antiangiogenics. *J Clin Oncol* 2001; 19: 1195-206.

[2] Kamen BA, Rubin E, Aisner J, et al. High-time chemotherapy or high time for low dose. *J Clin Oncol* 2000; 18: 2935-7.

[3] Hanahan D, Bergers G, Bergsland E. Less is more, regularly: metronomic dosing of cytotoxic drugs can target tumor angiogenesis in mice. *J Clin Invest* 2000; 105: 1045-7.

[4] Gasparini G. Metronomic scheduling: the future of chemotherapy? *Lancet Oncol* 2001; 2: 733-40.

[5] Skipper HE, Schabel FM, Jr., Mellett LB, et al. Implications of biochemical, cytokinetic, pharmacologic, and toxicologic relationships in the design of optimal therapeutic schedules. *Cancer Chemother Rep* 1970; 54: 431-50.

[6] DeVita VJ. (1997). *Principles of cancer management: Chemtherapy*. Philadelphia, Lippincott-Raven.

[7] Norton L. Implications of kinetic heterogeneity in clinical oncology. *Semin Oncol* 1985; 12: 231-49.

[8] Norton L. Evolving concepts in the systemic drug therapy of breast cancer. *Semin Oncol* 1997; 24: S10-3-S-.

[9] Seidman AD, Hudis CA, Albanel J, et al. Dose-dense therapy with weekly 1-hour paclitaxel infusions in the treatment of metastatic breast cancer. *J Clin Oncol* 1998; 16: 3353-61.

[10] Citron ML, Berry DA, Cirrincione C, et al. Randomized trial of dose-dense versus conventionally scheduled and sequential versus concurrent combination chemotherapy as postoperative adjuvant treatment of node-positive primary breast cancer: first report

of Intergroup Trial C9741/Cancer and Leukemia Group B Trial 9741. *J Clin Oncol* 2003; 21: 1431-9.

[11] Folkman J. Tumor angiogenesis: therapeutic implications. *N Engl J Med* 1971; 285: 1182-6.

[12] Hanahan D, Weinberg RA. The hallmarks of cancer. *Cell* 2000; 100: 57-70.

[13] Hafner C, Reichle A, Vogt T. New indications for established drugs: combined tumor-stroma-targeted cancer therapy with PPARgamma agonists, COX-2 inhibitors, mTOR antagonists and metronomic chemotherapy. *Curr Cancer Drug Targets* 2005; 5: 393-419.

[14] Weitman SD, Glatstein E, Kamen BA. Back to the basics: the importance of concentration x time in oncology. *J Clin Oncol* 1993; 11: 820-1.

[15] Eberhard A, Kahlert S, Goede V, et al. Heterogeneity of angiogenesis and blood vessel maturation in human tumors: implications for antiangiogenic tumor therapies. *Cancer Res* 2000; 60: 1388-93.

[16] Browder T, Butterfield CE, Kraling BM, et al. Antiangiogenic scheduling of chemotherapy improves efficacy against experimental drug-resistant cancer. *Cancer Res* 2000; 60: 1878-86.

[17] DiPaola RS, Durivage HJ, Kamen BA. High time for low-dose prospective clinical trials. *Cancer* 2003; 98: 1559-61.

[18] Crawford J, Ozer H, Stoller R, et al. Reduction by granulocyte colony-stimulating factor of fever and neutropenia induced by chemotherapy in patients with small-cell lung cancer. *N Engl J Med* 1991; 325: 164-70.

[19] Hamano Y, Sugimoto H, Soubasakos MA, et al. Thrombospondin-1 associated with tumor microenvironment contributes to low-dose cyclophosphamide-mediated endothelial cell apoptosis and tumor growth suppression. *Cancer Res* 2004; 64: 1570-4.

[20] Bocci G, Francia G, Man S, et al. Thrombospondin 1, a mediator of the antiangiogenic effects of low-dose metronomic chemotherapy. *Proc Natl Acad Sci U S A* 2003; 100: 12917-22.

[21] Damber JE, Vallbo C, Albertsson P, et al. The anti-tumour effect of low-dose continuous chemotherapy may partly be mediated by thrombospondin. *Cancer Chemother Pharmacol* 2005: 1-7.

[22] Dawson DW, Pearce SF, Zhong R, et al. CD36 mediates the In vitro inhibitory effects of thrombospondin-1 on endothelial cells. *J Cell Biol* 1997; 138: 707-17.

[23] Volpert OV, Zaichuk T, Zhou W, et al. Inducer-stimulated Fas targets activated endothelium for destruction by anti-angiogenic thrombospondin-1 and pigment epithelium-derived factor. *Nat Med* 2002; 8: 349-57.

[24] Quesada AJ, Nelius T, Yap R, et al. In vivo upregulation of CD95 and CD95L causes synergistic inhibition of angiogenesis by TSP1 peptide and metronomic doxorubicin treatment. *Cell Death Differ* 2005; 12: 649-58.

[25] Yap R, Veliceasa D, Emmenegger U, et al. Metronomic low-dose chemotherapy boosts CD95-dependent antiangiogenic effect of the thrombospondin peptide ABT-510: a complementation antiangiogenic strategy. *Clin Cancer Res* 2005; 11: 6678-85.

[26] Bocci G, Nicolaou KC, Kerbel RS. Protracted low-dose effects on human endothelial cell proliferation and survival in vitro reveal a selective antiangiogenic window for various chemotherapeutic drugs. *Cancer Res* 2002; 62: 6938-43.

[27] Kerbel RS, Kamen BA. The anti-angiogenic basis of metronomic chemotherapy. *Nat Rev Cancer* 2004; 4: 423-36.

[28] Garcia-Barros M, Paris F, Cordon-Cardo C, et al. Tumor response to radiotherapy regulated by endothelial cell apoptosis. *Science* 2003; 300: 1155-9.

[29] Slaton JW, Perrotte P, Inoue K, et al. Interferon-alpha-mediated down-regulation of angiogenesis-related genes and therapy of bladder cancer are dependent on optimization of biological dose and schedule. *Clin Cancer Res* 1999; 5: 2726-34.

[30] Ezekowitz RA, Mulliken JB, Folkman J. Interferon alfa-2a therapy for life-threatening hemangiomas of infancy. *N Engl J Med* 1992; 326: 1456-63.

[31] Kaban LB, Mulliken JB, Ezekowitz RA, et al. Antiangiogenic therapy of a recurrent giant cell tumor of the mandible with interferon alfa-2a. *Pediatrics* 1999; 103: 1145-9.

[32] Bisland SK, Lilge L, Lin A, et al. Metronomic photodynamic therapy as a new paradigm for photodynamic therapy: rationale and preclinical evaluation of technical feasibility for treating malignant brain tumors. *Photochem Photobiol* 2004; 80: 22-30.

[33] Bogaards A, Varma A, Zhang K, et al. Fluorescence image-guided brain tumour resection with adjuvant metronomic photodynamic therapy: pre-clinical model and technology development. *Photochem Photobiol Sci* 2005; 4: 438-42.

[34] Hahnfeldt P, Folkman J, Hlatky L. Minimizing long-term tumor burden: the logic for metronomic chemotherapeutic dosing and its antiangiogenic basis. *J Theor Biol* 2003; 220: 545-54.

[35] Stoll BR, Migliorini C, Kadambi A, et al. A mathematical model of the contribution of endothelial progenitor cells to angiogenesis in tumors: implications for antiangiogenic therapy. *Blood* 2003; 102: 2555-61.

[36] Asahara T, Murohara T, Sullivan A, et al. Isolation of putative progenitor endothelial cells for angiogenesis. *Science* 1997; 275: 964-7.

[37] Rafii S, Lyden D, Benezra R, et al. Vascular and haematopoietic stem cells: novel targets for anti-angiogenesis therapy? *Nat Rev Cancer* 2002; 2: 826-35.

[38] Bertolini F, Paul S, Mancuso P, et al. Maximum tolerable dose and low-dose metronomic chemotherapy have opposite effects on the mobilization and viability of circulating endothelial progenitor cells. *Cancer Res* 2003; 63: 4342-6.

[39] Asahara T, Masuda H, Takahashi T, et al. Bone marrow origin of endothelial progenitor cells responsible for postnatal vasculogenesis in physiological and pathological neovascularization. *Circ Res* 1999; 85: 221-8.

[40] Monestiroli S, Mancuso P, Burlini A, et al. Kinetics and viability of circulating endothelial cells as surrogate angiogenesis marker in an animal model of human lymphoma. *Cancer Res* 2001; 61: 4341-4.

[41] Tran J, Master Z, Yu JL, et al. A role for survivin in chemoresistance of endothelial cells mediated by VEGF. *Proc Natl Acad Sci U S A* 2002; 99: 4349-54.

[42] Toi M, Matsumoto T, Bando H. Vascular endothelial growth factor: its prognostic, predictive, and therapeutic implications. *Lancet Oncol* 2001; 2: 667-73.

[43] Toi M, Atiqur Rahman M, Bando H, et al. Thymidine phosphorylase (platelet-derived endothelial-cell growth factor) in cancer biology and treatment. *Lancet Oncol* 2005; 6: 158-66.

[44] Hurwitz H, Fehrenbacher L, Novotny W, et al. Bevacizumab plus irinotecan, fluorouracil, and leucovorin for metastatic colorectal cancer. *N Engl J Med* 2004; 350: 2335-42.

[45] Norton L, Simon R, Brereton HD, et al. Predicting the course of Gompertzian growth. *Nature* 1976; 264: 542-5.

[46] Klement G, Baruchel S, Rak J, et al. Continuous low-dose therapy with vinblastine and VEGF receptor-2 antibody induces sustained tumor regression without overt toxicity. *J Clin Invest* 2000; 105: R15-24.

[47] Klement G, Huang P, Mayer B, et al. Differences in therapeutic indexes of combination metronomic chemotherapy and an anti-VEGFR-2 antibody in multidrug-resistant human breast cancer xenografts. *Clin Cancer Res* 2002; 8: 221-32.

[48] Man S, Bocci G, Francia G, et al. Antitumor effects in mice of low-dose (metronomic) cyclophosphamide administered continuously through the drinking water. *Cancer Res* 2002; 62: 2731-5.

[49] Takahashi N, Haba A, Matsuno F, et al. Antiangiogenic therapy of established tumors in human skin/severe combined immunodeficiency mouse chimeras by anti-endoglin (CD105) monoclonal antibodies, and synergy between anti-endoglin antibody and cyclophosphamide. *Cancer Res* 2001; 61: 7846-54.

[50] Rozados VR, Sanchez AM, Gervasoni SI, et al. Metronomic therapy with cyclophosphamide induces rat lymphoma and sarcoma regression, and is devoid of toxicity. *Ann Oncol* 2004; 15: 1543-50.

[51] Bergers G, Hanahan D. Combining antiangiogenic agents with metronomic chemotherapy enhances efficacy against late-stage pancreatic islet carcinomas in mice. *Cold Spring Harb Symp Quant Biol* 2002; 67: 293-300.

[52] Bello L, Carrabba G, Giussani C, et al. Low-dose chemotherapy combined with an antiangiogenic drug reduces human glioma growth in vivo. *Cancer Res* 2001; 61: 7501-6.

[53] Zhao D, Jiang L, Hahn EW, et al. Continuous low-dose (metronomic) chemotherapy on rat prostate tumors evaluated using MRI in vivo and comparison with histology. *Neoplasia* 2005; 7: 678-87.

[54] Soffer SZ, Moore JT, Kim E, et al. Combination antiangiogenic therapy: increased efficacy in a murine model of Wilms tumor. *J Pediatr Surg* 2001; 36: 1177-81.

[55] Abraham D, Abri S, Hofmann M, et al. Low dose carboplatin combined with angiostatic agents prevents metastasis in human testicular germ cell tumor xenografts. *J Urol* 2003; 170: 1388-93.

[56] Zhang L, Yu D, Hicklin DJ, et al. Combined anti-fetal liver kinase 1 monoclonal antibody and continuous low-dose doxorubicin inhibits angiogenesis and growth of human soft tissue sarcoma xenografts by induction of endothelial cell apoptosis. *Cancer Res* 2002; 62: 2034-42.

[57] Klink T, Bela C, Stoelting S, et al. Metronomic trofosfamide inhibits progression of human lung cancer xenografts by exerting anti-angiogenic effects. *J Cancer Res Clin Oncol* 2006.

[58] Hermans IF, Chong TW, Palmowski MJ, et al. Synergistic effect of metronomic dosing of cyclophosphamide combined with specific antitumor immunotherapy in a murine melanoma model. *Cancer Res* 2003; 63: 8408-13.

[59] Pietras K, Hanahan D. A multitargeted, metronomic, and maximum-tolerated dose "chemo-switch" regimen is antiangiogenic, producing objective responses and survival benefit in a mouse model of cancer. *J Clin Oncol* 2005; 23: 939-52.

[60] Emmenegger U, Morton GC, Francia G, et al. Low-dose metronomic daily cyclophosphamide and weekly tirapazamine: a well-tolerated combination regimen with enhanced efficacy that exploits tumor hypoxia. *Cancer Res* 2006; 66: 1664-74.

[61] Sumiyoshi K, Strebel FR, Rowe RW, et al. The effect of whole-body hyperthermia combined with 'metronomic' chemotherapy on rat mammary adenocarcinoma metastases. *Int J Hyperthermia* 2003; 19: 103-18.

[62] Munoz R, Man S, Shaked Y, et al. Highly efficacious nontoxic preclinical treatment for advanced metastatic breast cancer using combination oral UFT-cyclophosphamide metronomic chemotherapy. *Cancer Res* 2006; 66: 3386-91.

[63] Ma L, Francia G, Viloria-Petit A, et al. In vitro procoagulant activity induced in endothelial cells by chemotherapy and antiangiogenic drug combinations: modulation by lower-dose chemotherapy. *Cancer Res* 2005; 65: 5365-73.

[64] Shaked Y, Emmenegger U, Francia G, et al. Low-dose metronomic combined with intermittent bolus-dose cyclophosphamide is an effective long-term chemotherapy treatment strategy. *Cancer Res* 2005; 65: 7045-51.

[65] Emmenegger U, Man S, Shaked Y, et al. A comparative analysis of low-dose metronomic cyclophosphamide reveals absent or low-grade toxicity on tissues highly sensitive to the toxic effects of maximum tolerated dose regimens. *Cancer Res* 2004; 64: 3994-4000.

[66] Duration and intensity of maintenance chemotherapy in acute lymphoblastic leukaemia: overview of 42 trials involving 12 000 randomised children. Childhood ALL Collaborative Group. *Lancet* 1996; 347: 1783-8.

[67] Joussen AM, Kruse FE, Volcker HE, et al. Topical application of methotrexate for inhibition of corneal angiogenesis. *Graefes Arch Clin Exp Ophthalmol* 1999; 237: 920-7.

[68] Presta M, Rusnati M, Belleri M, et al. Purine analogue 6-methylmercaptopurine riboside inhibits early and late phases of the angiogenesis process. *Cancer Res* 1999; 59: 2417-24.

[69] Hirata S, Matsubara T, Saura R, et al. Inhibition of in vitro vascular endothelial cell proliferation and in vivo neovascularization by low-dose methotrexate. *Arthritis Rheum* 1989; 32: 1065-73.

[70] Rubie H, Coze C, Plantaz D, et al. Localised and unresectable neuroblastoma in infants: excellent outcome with low-dose primary chemotherapy. *Br J Cancer* 2003; 89: 1605-9.

[71] Crist WM, Anderson JR, Meza JL, et al. Intergroup rhabdomyosarcoma study-IV: results for patients with nonmetastatic disease. *J Clin Oncol* 2001; 19: 3091-102.

[72] Pazdur R, Lassere Y, Rhodes V, et al. Phase II trial of uracil and tegafur plus oral leucovorin: an effective oral regimen in the treatment of metastatic colorectal carcinoma. *J Clin Oncol* 1994; 12: 2296-300.

[73] Ota K, Taguchi T, Kimura K. Report on nationwide pooled data and cohort investigation in UFT phase II study. *Cancer Chemother Pharmacol* 1988; 22: 333-8.

[74] Sakamoto J, Ohashi Y, Hamada C, et al. Efficacy of oral adjuvant therapy after resection of colorectal cancer: 5-year results from three randomized trials. *J Clin Oncol* 2004; 22: 484-92.

[75] Lembersky BC, Wieand HS, Petrelli NJ, et al. Oral uracil and tegafur plus leucovorin compared with intravenous fluorouracil and leucovorin in stage II and III carcinoma of the colon: results from National Surgical Adjuvant Breast and Bowel Project Protocol C-06. *J Clin Oncol* 2006; 24: 2059-64.

[76] Noguchi S, Koyama H, Uchino J, et al. Postoperative adjuvant therapy with tamoxifen, tegafur plus uracil, or both in women with node-negative breast cancer: a pooled analysis of six randomized controlled trials. *J Clin Oncol* 2005; 23: 2172-84.

[77] Fennelly D, Aghajanian C, Shapiro F, et al. Phase I and pharmacologic study of paclitaxel administered weekly in patients with relapsed ovarian cancer. *J Clin Oncol* 1997; 15: 187-92.

[78] Greco FA. Docetaxel (Taxotere) administered in weekly schedules. *Semin Oncol* 1999; 26: 28-31.

[79] Burstein HJ, Manola J, Younger J, et al. Docetaxel administered on a weekly basis for metastatic breast cancer. *J Clin Oncol* 2000; 18: 1212-9.

[80] Kuroi K, Bando H, Nagai S, et al. Efficacy of weekly docetaxel therapy for advanced or recurrent breast cancer. *Jpn J Cancer Chemother* 2001; 28: 797-802.

[81] Kuroi K, Bando H, Saji S, et al. Protracted administration of weekly docetaxel in metastatic breast cancer. *Oncol Rep* 2003; 10: 1479-84.

[82] Kuroi K, Toi M. Current perspectives of weekly schedule of docetaxel in breast cancer. http://www.siicsalud.com/dato/dat046/06109003.htm 2006.

[83] Gately S, Kerbel R. Antiangiogenic scheduling of lower dose cancer chemotherapy. *Cancer J* 2001; 7: 427-36.

[84] Kakolyris S, Samonis G, Koukourakis M, et al. Treatment of non-small-cell lung cancer with prolonged oral etoposide. *Am J Clin Oncol* 1998; 21: 505-8.

[85] Niitsu N, Umeda M. Evaluation of long-term daily administration of oral low-dose etoposide in elderly patients with relapsing or refractory non-Hodgkin's lymphoma. *Am J Clin Oncol* 1997; 20: 311-4.

[86] Braybrooke JP, O'Byrne KJ, Propper DJ, et al. A phase II study of razoxane, an antiangiogenic topoisomerase II inhibitor, in renal cell cancer with assessment of potential surrogate markers of angiogenesis. *Clin Cancer Res* 2000; 6: 4697-704.

[87] Sterba J, Pavelka Z, Slampa P. Concomitant radiotherapy and metronomic temozolomide in pediatric high-risk brain tumors. *Neoplasma* 2002; 49: 117-20.

[88] Danson S, Lorigan P, Arance A, et al. Randomized phase II study of temozolomide given every 8 hours or daily with either interferon alfa-2b or thalidomide in metastatic malignant melanoma. *J Clin Oncol* 2003; 21: 2551-7.

[89] Hwu WJ, Krown SE, Menell JH, et al. Phase II study of temozolomide plus thalidomide for the treatment of metastatic melanoma. *J Clin Oncol* 2003; 21: 3351-6.

[90] Keldsen N, Madsen EL, Havsteen H, et al. Oral treosulfan as second-line treatment in platinum-resistant ovarian cancer: a phase II study. The Danish Ovarian Cancer Study Group. *Gynecol Oncol* 1998; 69: 100-2.

[91] Gunsilius E, Gierlich T, Mross K, et al. Palliative chemotherapy in pretreated patients with advanced cancer: oral trofosfamide is effective in ovarian carcinoma. *Cancer Invest* 2001; 19: 808-11.

[92] Azzarelli A, Gronchi A, Bertulli R, et al. Low-dose chemotherapy with methotrexate and vinblastine for patients with advanced aggressive fibromatosis. *Cancer* 2001; 92: 1259-64.

[93] Colleoni M, Rocca A, Sandri MT, et al. Low-dose oral methotrexate and cyclophosphamide in metastatic breast cancer: antitumor activity and correlation with vascular endothelial growth factor levels. *Ann Oncol* 2002; 13: 73-80.

[94] Glode LM, Barqawi A, Crighton F, et al. Metronomic therapy with cyclophosphamide and dexamethasone for prostate carcinoma. *Cancer* 2003; 98: 1643-8.

[95] Wolff JE, Molenkamp G, Hotfilder M, et al. Dexamethasone inhibits glioma-induced formation of capillary like structures in vitro and angiogenesis in vivo. *Klin Padiatr* 1997; 209: 275-7.

[96] Machein MR, Kullmer J, Ronicke V, et al. Differential downregulation of vascular endothelial growth factor by dexamethasone in normoxic and hypoxic rat glioma cells. *Neuropathol Appl Neurobiol* 1999; 25: 104-12.

[97] Spieth K, Kaufmann R, Gille J. Metronomic oral low-dose treosulfan chemotherapy combined with cyclooxygenase-2 inhibitor in pretreated advanced melanoma: a pilot study. *Cancer Chemother Pharmacol* 2003; 52: 377-82.

[98] Vogt T, Hafner C, Bross K, et al. Antiangiogenetic therapy with pioglitazone, rofecoxib, and metronomic trofosfamide in patients with advanced malignant vascular tumors. *Cancer* 2003; 98: 2251-6.

[99] Coras B, Hafner C, Reichle A, et al. Antiangiogenic therapy with pioglitazone, rofecoxib, and trofosfamide in a patient with endemic kaposi sarcoma. *Arch Dermatol* 2004; 140: 1504-7.

[100] Nicolini A, Mancini P, Ferrari P, et al. Oral low-dose cyclophosphamide in metastatic hormone refractory prostate cancer (MHRPC). *Biomed Pharmacother* 2004; 58: 447-50.

[101] Reichle A, Bross K, Vogt T, et al. Pioglitazone and rofecoxib combined with angiostatically scheduled trofosfamide in the treatment of far-advanced melanoma and soft tissue sarcoma. *Cancer* 2004; 101: 2247-56.

[102] Fassas AB, Kiwan E, Roberts J, et al. Metronomic chemoimmunotherapy in a patient with refractory Waldenstrom's macroglobulinemia. *Leuk Lymphoma* 2005; 46: 1675-7.

[103] Herrlinger U, Rieger J, Steinbach JP, et al. UKT-04 trial of continuous metronomic low-dose chemotherapy with methotrexate and cyclophosphamide for recurrent glioblastoma. *J Neurooncol* 2005; 71: 295-9.
[104] Kieran MW, Turner CD, Rubin JB, et al. A feasibility trial of antiangiogenic (metronomic) chemotherapy in pediatric patients with recurrent or progressive cancer. *J Pediatr Hematol Oncol* 2005; 27: 573-81.
[105] Bocci G, Tuccori M, Emmenegger U, et al. Cyclophosphamide-methotrexate 'metronomic' chemotherapy for the palliative treatment of metastatic breast cancer. A comparative pharmacoeconomic evaluation. *Ann Oncol* 2005; 16: 1243-52.
[106] Gille J, Spieth K, Kaufmann R. Metronomic low-dose chemotherapy as antiangiogenic therapeutic strategy for cancer. *J Dtsch Dermatol Ges* 2005; 3: 26-32.
[107] Polverini PJ ,Novak RF. Inhibition of angiogenesis by the antineoplastic agents mitoxantrone and bisantrene. *Biochem Biophys Res Commun* 1986; 140: 901-7.
[108] Albertsson P, Lennernas B, Norrby K. On metronomic chemotherapy: Modulation of angiogenesis mediated by VEGE-A. *Acta Oncol* 2006; 45: 144-55.
[109] Poon RT, Fan ST, Wong J. Clinical implications of circulating angiogenic factors in cancer patients. *J Clin Oncol* 2001; 19: 1207-25.
[110] Kuroi K, Toi M. Circulating angiogenesis regulators in cancer patients. *Int J Biol Markers* 2001; 16: 5-26.
[111] Kuroi K, Tanaka C, Toi M. Circulating levels of endostatin in cancer patients. *Oncol Rep* 2001; 8: 405-9.
[112] Maraveyas A, Lam T, Hetherington JW, et al. Can a rational design for metronomic chemotherapy dosing be devised? *Br J Cancer* 2005; 92: 1588-90.
[113] Lam T, Hetherington JW, Greenman J, et al. From total empiricism to a rational design of metronomic chemotherapy phase I dosing trials. *Anticancer Drugs* 2006; 17: 113-21.
[114] Shaked Y, Bertolini F, Man S, et al. Genetic heterogeneity of the vasculogenic phenotype parallels angiogenesis; Implications for cellular surrogate marker analysis of antiangiogenesis. *Cancer Cell* 2005; 7: 101-11.
[115] Shaked Y, Emmenegger U, Man S, et al. Optimal biologic dose of metronomic chemotherapy regimens is associated with maximum antiangiogenic activity. *Blood* 2005; 106: 3058-61.
[116] Mancuso P, Calleri A, Cassi C, et al. Circulating endothelial cells as a novel marker of angiogenesis. *Adv Exp Med Biol* 2003; 522: 83-97.
[117] Mancuso P, Colleoni M, Calleri A, et al. Circulating endothelial cell kinetics and viability predict survival in breast cancer patients receiving metronomic chemotherapy. *Blood* 2006.
[118] Schneider M, Tjwa M, Carmeliet P. A surrogate marker to monitor angiogenesis at last. *Cancer Cell* 2005; 7: 3-4.
[119] Goon PK, Lip GY, Boos CJ, et al. Circulating endothelial cells, endothelial progenitor cells, and endothelial microparticles in cancer. *Neoplasia* 2006; 8: 79-88.
[120] Buteau-Lozano H, Ancelin M, Lardeux B, et al. Transcriptional regulation of vascular endothelial growth factor by estradiol and tamoxifen in breast cancer cells: a complex interplay between estrogen receptors alpha and beta. *Cancer Res* 2002; 62: 4977-84.

[121] Higgins KJ, Liu S, Abdelrahim M, et al. Vascular endothelial growth factor receptor-2 expression is induced by 17beta-estradiol in ZR-75 breast cancer cells by estrogen receptor alpha/Sp proteins. *Endocrinology* 2006; 147: 3285-95.

In: New Research on Angiogenesis Inhibitors
Editor: Elmer T. Skinard, pp. 77-98
ISBN: 1-60021-473-8

Chapter IV

COMBINED RADIOIMMUNOTHERAPY AND ANTIANGIOGENIC THERAPY TO TREAT CANCER: VALIDATION STUDIES IN ANIMAL MODELS

Seigo Kinuya*
Department of Biotracer Medicine (formerly Department of Nuclear Medicine),
Kanazawa University Graduate School of Medical Sciences, Kanazawa, Japan.

ABSTRACT

Radioimmunotherapy (RIT) employing radiolabeled monoclonal antibodies (MAb) against tumors has been extensively tested for more than two decades. However, in most cases, the results have demonstrated limited success in bulky disease. In contrast, several reports suggest the role of RIT as an adjuvant treatment of cancer. Neovascularization is a prerequisite for the growth of solid tumors; therefore, antiangiogenic therapy employing agents that suppress endothelial cell growth has been investigated. Antiangiogenic therapy does not generally cause bone marrow suppression or gastrointestinal symptoms. However, antiangiogenic therapy is inhibitory or cytostatic only, which may maintain lesions at limited size, a situation referred to as tumor dormancy. Therefore, combination with cytotoxic treatment is required for the extermination of tumor cells. We hypothesize that the combined utility of antiangiogenic therapy and RIT may cooperatively effect treatment of cancer lesions. The rationale for this hypothesis is based on suppression of tumor endothelial compartments by antiangiogenic agents as well as the direct action of RIT against tumor cells. Additionally, this combination may not increase bone marrow toxicity, which is the major determinant of maximal tolerated dose of RIT. This strategy was examined in a subcutaneous xenograft model and a liver metastasis model of human colon cancer cells

* Correspondence concerning this article should be addressed to Seigo Kinuya, M.D., Ph.D., Department of Biotracer Medicine, Kanazawa University Graduate School of Medical Sciences, 13-1 Takaramachi, Kanazawa, Ishikawa 920-8640, Japan. Tel:+81-76-265-2333; Fax:+81-76-234-4257; E-mail: kinuya@med.kanazawa-u.ac.jp.

using ^{131}I-labeled anti-colorectal monoclonal antibody and two agents that are known to exhibit antiangiogenic property, 2-methoxyestradiol and thalidomide.

INTRODUCTION

Paul Ehrlich hypothesized a century ago that a magic bullet could be developed to selectively target disease. The establishment of hybridoma technology by Kohler and Milstein made this vision practical by providing monoclonal antibodies (MAb) that possess highly specific binding capacity to their targeted antigens [1]. Currently, a number of MAbs against various types of malignancies have been approved in the United States and other counties (Table 1). Many of them use mechanism such as antibody-dependent cellular cytotoxicity and complement-dependent cytotoxicity in effecting cytotoxicity to tumor cells [2]. MAbs can also be used to target payloads to directly kill tumor cells, including radionuclides, drugs and toxins.

Table 1. Therapeutic monoclonal antibodies.

Name (Trade name)	Origin	Isotype	Target		FDA approval
Unconjugates					
Trastuzumab (Herception)	Humanized	Human IgG1	HER2/neu	Breast cancer	1998
Rituximab (Rituxan)	Chimeric	Human IgG1	CD20	Lymphoma	1997
Cetuximab (Erbitux)	Chimeric	Human IgG1	EGFR	Colorectal caner	2004
Bevacizumab (Avastin)	Chimeric	Human IgG1	VEGF	Colorectal and lung cancers	2004
Alemtuzumab (Campath-1H)	Humanized	Human IgG1	CD52	Chronic lymphocytic leukemia	2001
Immunoconjugates					
Ibritumomab tiuxetan (Zevalin)	Murine	Murine IgG1 labeled with Y-90	CD20	Lymphoma	2002
Tositumomab (Bexxar)	Murine	Murine IgG2a labeled with I-131	CD20	Lymphoma	2003
Gemtuzumab (Myelotarg)	Human	Human IgG4 conjugated to calicheamicin	CD33	Acute myelogenous leukemia	2000

Radioimmunotherapy (RIT) employing radiolabeled MAb has been extensively tested for more than two decades [3]. Clinical results of RIT in patients with malignant lymphoma are encouraging, including ^{90}Y-labaled murine IgG1 against CD20 cell surface antigen on B-cell lymphoma, Ibritumobab tiuxetan (Zevalin®) and ^{131}I-labeld IgG2a against the same antigen, Tositumomab (Bexxar®) [2-4]. However, in solid tumors, RIT with a single systemic injection of radiolabeled MAb produces insufficient tumor response even at its maximum tolerated administration dose [2-4]. Overall response rates are often only less than 10% [3].

One of the major reasons of unsatisfactory results of RIT in solid tumors would involve limited targeting of radiolabeled MAb to tumors as well as low sensitivity to radiation of them [4]: <0.01% of the injected dose localizes in one gram of tumor, which delivers only a cumulative dose of less than 1,500 cGy whereas >5,000 cGy is usually needed to achieve a therapeutic response in most malignancies [5]. Several barriers exist for the targeting of radiation in tumors, including heterogeneous blood supply and transport across interstitial distances [5]. Therefore, several strategies have been developed, including pre-targeting the system for better tumor-to-nontumor uptake ratios [3], integration of bone marrow support to allow delivery of higher radiation doses to tumors [6,7], use of combinations of RIT and chemotherapy [8-10] or hyperthermia [11,12] and the utility of biological response modifiers for more accurate delivery of labeled MAb [13-15].

In most cases, the results of clinical RIT have demonstrated limited success in bulky disease. In contrast, several reports suggest the role of RIT as an adjuvant treatment of cancer where small residual tumors are better and more homogeneously targeted by radiolabeled MAbs than large tumors are [4,16-27]. Indeed, prolonged survival with RIT has been documented in patients presenting with minimal residual disease of ovarian cancer [28].

Neovascularization is a prerequisite for the growth of solid tumors [29-32]. Tumor cells produce angiogenic factors, such as vascular endothelial growth factor (VEGF), which induce neovascularization to support their own growth [29-32]; therefore, antiangiogenic therapy employing agents that suppress endothelial cell growth has been investigated. Moreover, clinical trials involving a number of angiogenesis inhibitors are underway. Antiangiogenic therapy affords several advantages over conventional chemotherapy [29]. Firstly, antiangiogenic therapy does not generally cause bone marrow suppression or gastrointestinal symptoms. Secondly, drug resistance most likely is not a significant problem. However, antiangiogenic therapy is inhibitory or cytostatic only, which may maintain lesions at limited size, a situation referred to as tumor dormancy. Therefore, combination with cytotoxic treatment is required for the extermination of tumor cells.

We hypothesize that the combined utility of antiangiogenic therapy and RIT may cooperatively effect treatment of cancer lesions. The rationale for this hypothesis is based on suppression of tumor endothelial compartments by antiangiogenic agents as well as the direct action of RIT against tumor cells. Additionally, this combination may not increase bone marrow toxicity, which is the major determinant of maximal tolerated dose (MTD) of RIT. We have examined this hypothesis in a subcutaneous xenograft model [33,34] and a liver metastasis model [24,35] of human colon cancer cells using ^{131}I-labeled anti-colorectal monoclonal antibody and two agents that are known to exhibit antiangiogenic property, 2-methoxyestradiol (2-ME) and thalidomide [29-32], demonstrating the synergism of RIT and antiangiogenic property. 2-ME is currently in being evaluated in clinical trials for the treatment of various types of tumors such as prostate cancer and breast cancer, in which 2-ME is reportedly well tolerated and demonstrate promising activity [32]. Thalidomide is also in varieties of clinical trials [32].

Materials and Methods

Monoclonal Antibodies and Radiolabeling

A7, an IgG1 murine monoclonal antibody (MAb) with *k*-light chain, was used. It was purified from ascites of hybridoma-bearing mice by Protein A Sepharose column (Bio-Rad, Richmond, CA), recognizing a 45 kD surface glycoprotein of human colonic carcinomas [36]. The antigen is modulated after the binding with A7, and the internalization of A7 occurs. A7 does not react with normal gastric mucosa, erythrocytes, peripheral lymphocytexs or ileal mucosa, and shows weak reactivity with 10% of colon mucosa specimens [36]. An additional IgG1 recognizing placental alkaline phosphatase, HPMS-1, was utilized as a class-matched irrelevant control MAb [37]. Antibodies were radiolabeled with ^{131}I or ^{125}I by the chloramine-T method, and were then purified utilizing a PD10 column (Pharmacia LKB Biotechnology, Uppsala, Sweden). The labeled MAbs were sterilized by filtration (Millex-GV, 0.22 μm; Millipore, Bedford, MA) prior to further experiments. The specific activity of the purified MAbs was approximately 111 MBq (3 mCi)/mg. In addition, the immunoreactivity of labeled A7 was about 70% at infinite antigen excess determined by reaction with LS180 human colon cancer cells (American Type Culture Collection, Rockville, MD). LS180 cells were grown in DMEM medium (Nissui Seiyaku, Tokyo, Japan) and harvested with 0.1% trypsin and 0.02% EDTA. This cell line reportedly shows considerable VEGF production; moreover, the growth of xenografts in mice is suppressed by an anti-VEGF antibody [38].

Animal Models

Animal studies were performed in compliance with the regulations of our institution. The principles of laboratory animal care (NIH publication No. 86-23, revised 1985) were followed. We prepared a subcutaneous xenograft model and a liver metastasis model of LS180 colon cancer cells. Briefly, Balb/c nu/nu mice (female, 20 g) were xenografted subcutaneously in the thigh area with 5×10^6 LS180 cells. These rodents were subsequently used when the xenografts reached approximately 5-6 mm in diameter. A liver metastasis model was established in Balb/c nu/nu mice by injecting 0.1 ml of the cell suspension at 1×10^7 into the spleen under inhalation anesthesia with diethyl ether [16,17,23]. The spleen was removed two minutes later. In this model, metastatic nodules of less than 1 mm in diameter are evident one week after cell inoculation [23]. At two weeks, many metastases attain sizes of several millimeters; moreover, these metastases often fuse with one another.

Antiangiogenic Therapy and in vitro Properties of the Agents

Antiangiogenic agents employed in this study were 2-methoxyestradiol (2-ME) (Sigma, St. Louis, MO) and thalidomide (Tocris, Ballwin, MO) [29-32]. Both agents were suspended in 0.5% carboxymethylcellulose (Wako, Osaka, Japan). A daily dose of 75 mg/kg of 2-ME

[39] or a dose of 200 mg/kg of thalidomide [40-42] was administered orally. These doses were determined base on previous reports [39-42].

In order to determine whether 2-ME and thalidomide would exert direct cytotoxicity on LS180 cells, the cells were incubated in 24-well culture plates at 20,000 cells/ml in DMEM medium (Nissui Seiyaku, Tokyo, Japan) with increasing concentrations of the agents. The agents were dissolved in dimethyl sulfoxide (DMSO; Wako, Tokyo, Japan) and diluted in the medium. The final concentration of DMSO was 0.5%. Control cells were treated with DMSO alone. Confirmation was obtained prior to the assay that the employed DMSO concentration did not affect proliferation of LS180 cells. Viable cell number was determined by tryptan blue exclusion 1, 3 and 5 days following inoculation (n=3).

It has been reported that cells exposed to 2-ME accumulate in the G2 phase of the cell cycle [43,44]. Cells are radiosensitive in G2 phase. Consequently, 2-ME may act as a radiosensitizer if cells are treated prior to irradiation [44]. Therefore, the possible radiosensitization effect of 2-ME on LS180 cells was assessed by clonogenic assay. Briefly, LS180 cells were treated with various concentrations of 2-ME for 24 h, followed by exposure to 0-4 Gy of an X-ray beam (MBR-1520R; Hitachi, Tokyo, Japan) at 1.41 Gy/min. The cells were washed and incubated in drug-free medium. Subsequently, cell survival was determined by counting colonies fixed with 80% methanol and stained with crystal violet. Colonies containing 50 or more cells were considered as survivors. Surviving fractions were corrected by plating efficiency at each concentration of 2-ME. Similarly, the possibility of radiosensitization effect of thalidomide was examined.

Validation in a Subcutaneous Model

Because the antiangiogenic influence may decrease the delivery of ^{131}I-A7 to xenografted tumors, the effect of 2-ME administration on ^{131}I-A7 distribution to the tumor was measured. In the first experiment, we tested whether 2-ME administration would affect the maximal tumor accumulation of ^{131}I-A7. A dose of 111 kBq of ^{131}I-A7 was injected intravenously into tumor-bearing mice (n=10) and the biodistribution was observed 2 days later when the maximum tumor uptake is achieved with this MAb and this animal model. A dose of 75 mg/kg of 2-ME was administered orally 1 h before the ^{131}I-A7 injection on day 0, day 1 and 1 h before the sacrifice of animal on day 2. In the second experiment, the effect of 2-ME on the retention of ^{131}I-A7 from the tumor was measured after maximal uptake had been achieved. In this design, the mice received 2-ME on day 2 and daily thereafter. The biodistribution of ^{131}I-A7 was observed on day 7 in this group (n=5).

Antitumor Effects of RIT with ^{131}I-A7 and antiangiogenic therapy with 2-ME were evaluated by measuring tumor volume (mm^3) that was calculated as length (mm) × width (mm^2) × 0.5. The value was expressed as a ratio of volumes to the volume obtained on day 0. The therapeutic experiment was comprised of several groups: non-treated mice, mice treated with antiangiogenic therapy consisting of a daily dose of 2-ME, mice treated with RIT of 4.63 MBq (125 μCi) ^{131}I-A7 or ^{131}I-HPMS-1 and mice treated with RIT combined with the antiangiogenic therapy (n=10). The dose of ^{131}I-A7 was approximately half of maximum tolerated dose in this model. Non-treated mice and mice treated with RIT alone were fed an identical volume of carboxymethylcellulose which did not contain 2-ME. For the combined treatment, the administration of 2-ME was initiated 2 days after the ^{131}I-A7 injection, based on the results of the biodistribution study. Animal body weights were monitored to assess any adverse effects of treatment. Similarly, effects of RIT with ^{131}I-A7 and another antiangiogenic therapy with thalidomide were observed. In the experiment with thalidomide, a blood sample (5 μl) was obtained from a tail vein in each mouse to ascertain the bone marrow toxicity. Samples from three mice selected at random within a single group were pooled and diluted 1:10000 in phosphate buffered saline for red blood cell (RBC) counts, 1:20 in 3% acetic acid for white blood cell (WBC) counts and 1:100 in 1% ammonium oxalate for platelet counts.

On day 15, tumors were excised from 2-ME-treated, thalidomide-treated and non-treated control rodents (n=3). Specimens were fixed with 10% formalin and embedded in paraffin wax. Immunoperoxidase staining was conducted with a rabbit polyclonal antibody against factor VIII (DAKO, Copenhagen, Denmark). Microvessel density was determined by light microscopy according to the procedure described by Weidner et al. [45]. Briefly, the areas of highest vascularization were found by scanning the tumor sections at low power (100×). Individual microvessels were subsequently counted on a 200× field. Five different fields were examined in each section. Two sections per tumor were studied and the averages of three tumors in each group were compared.

Validation in a Liver Metastasis Model

Accumulation of ^{125}I-A7 in liver metastases was examined. Mice were intravenously injected with 111 kBq (3 μCi) of ^{125}I-A7 two weeks after inoculation of tumor cells. The animals were sacrificed two (n=10) and six days (n=8) later. Organs were excised and weighed; subsequently, radioactivity of the tissues was measured with a well-type gamma counter. To determine ^{125}I-A7 accumulation in metastases, all visible nodules were separated from the liver. Total weight and radioactivity of nodules were then measured. Accumulation of the irrelevant MAb, ^{131}I-HMPS-1, was also assessed (n=4). In order to evaluate the effect of 2-ME administration on ^{125}I-A7 accumulation into metastases, a dose of 2-ME was administered orally one week after the cell inoculation and daily thereafter. ^{125}I-A7 was injected two weeks after the cell inoculation; its biodistribution was assessed two days later (n=5). The results were compared with those obtained from non-treated mice.

The therapeutic experiment consisted of six groups: non-treated mice (n=6), mice receiving antiangiogenic therapy with a daily oral dose of 75 mg/kg of 2-ME (n=8), mice undergoing specific RIT with 7 MBq (190 μCi) of ^{131}I-A7 (n=8), mice receiving both ^{131}I-A7 RIT and antiangiogenic therapy (n=8), mice undergoing non-specific RIT with 7 MBq of an irrelevant ^{131}I-HPMS-1 (n=6) and mice receiving both ^{131}I-HPMS-1 RIT and antiangiogenic therapy (n=8). To mimic the theoretical situation in which RIT would be performed during antiangiogenic therapy, antiangiogenic therapy was initiated one week following cell inoculation; RIT was performed at two weeks. ^{131}I-A7 dosage was determined based on a preliminary study involving 9 MBq (242 μCi) of ^{131}I-A7 administration in this liver metastasis model, in which half of the mice died due to radiation toxicity. The animals were sacrificed four weeks following the cell inoculation. Therapeutic effect was assessed by total weight of metastases.

In the next experiment, survival of animals was observed. This experiment consisted of six groups: non-treated mice (n=5), mice receiving antiangiogenic therapy consisting of a daily oral dose of 2-ME (n=8), mice undergoing specific RIT with 7 MBq (190 μCi) of ^{131}I-A7 (n=8), mice receiving both ^{131}I-A7 RIT and antiangiogenic therapy (n=10), mice undergoing non-specific RIT with 7 MBq of an irrelevant ^{131}I-HPMS-1 (n=5) and mice subjected to both ^{131}I-HPMS-1 RIT and antiangiogenic therapy (n=5). 2-ME administration was initiated three days following cell inoculation, and was continued daily thereafter for four weeks; RIT was performed at seven days. The results were analyzed by the Kaplan-Meier method with the Logrank test. The level of significance was set at 5%. To ascertain the bone marrow toxicity associated with this treatment, peripheral blood cell counts were monitored.

Results

In vitro Properties of Agents

In vitro proliferation assay demonstrated 2-ME dose-dependent cytotoxicity for LS180 cells at concentrations of 5 μmol/L and larger (Fig. 1). In contrast, the assay revealed no evidence of direct cytotoxicity of thalidomide in LS180 cells up to concentrations of 100 μmol/L (Fig. 2). Treatment with either 2-ME or thalidomide prior to x-ray exposure did not affect the surviving fraction of LS180 cells corrected by plating efficiency at each concentration of the agents (Fig. 3).

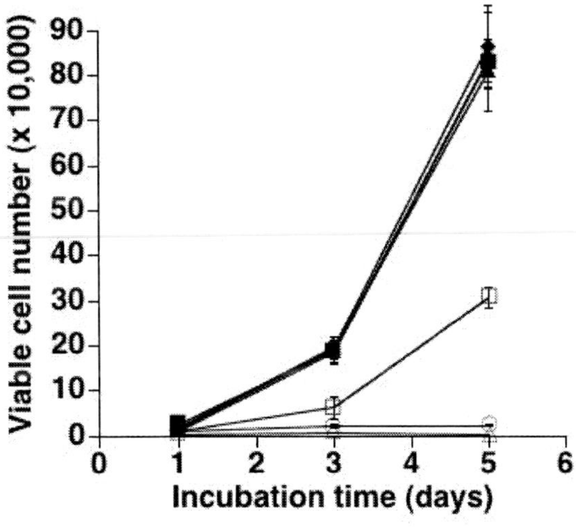

Fig. 1. *In vitro* proliferation assay of LS180 colon cancer cells in the presence of increasing concentrations of 2-methoxyestradiol (n=3). B, Control cells without 2-methoxyestradiol; J, cells incubated with 0.1 μM; H, 1 μM; F, 2 μM; G, 5 μM; E, 10 μM; C, 50 μM.

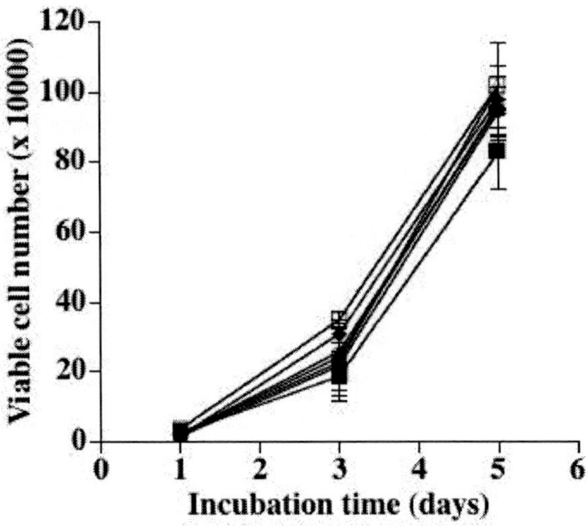

Fig. 2. In vitro proliferation assay of LS180 colon cancer cells in the presence of increasing concentrations of thalidomide (n=3). B, Without thalidomide; J, in the presence of thalidomide 0.1 µmol/L; H, 1 µmol/L; F, 5 µmol/L; G, 10 µmol/L; E, 50 µmol/L; C, 100 µmol/L.

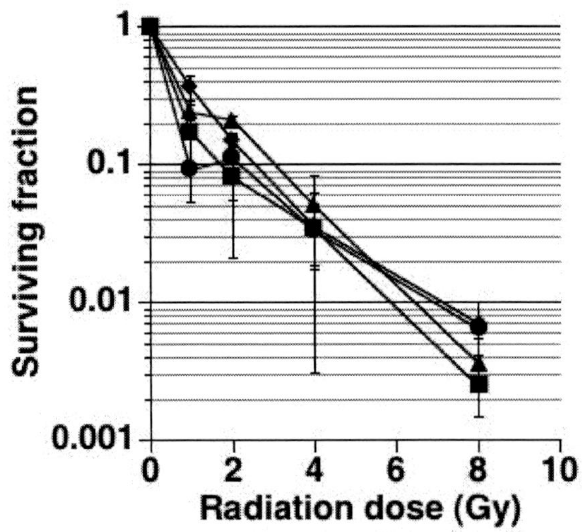

Fig. 3. Clonogenic assay of LS180 colon cancer cells exposed to x-ray radiation (n=3). Cells were incubated with 2-methoxyestradiol for 24 h prior to irradiation. Cells were then washed and incubated in drug-free medium. Cell survival was determined by counting colonies. Surviving fraction was corrected by plating efficiency at each concentration of 2-methoxyestradiol. B, Control cells; J, cells pretreated with 2-methoxyestradiol 1 µM for 24 h; H, 2-methoxyestradiol 2 µM; F, 2-methoxyestradiol 5 µM.

In a Subcutaneous Model

^{131}I-A7 accumulated almost exclusively in the LS180 xenografts, demonstrating activity of 29.4 ± 9.32 %ID/g 2 days after the injection in the control group. The administration of 2-

ME on days 0, 1 and 2 reduced the tumor uptake of ^{131}I-A7 to 20.0± 5.61 %ID/g ($p < 0.005$) without affecting the distribution of ^{131}I-A7 in normal tissues (Fig. 4A). In the second experiment of the biodistribution study, 2-ME was administered on day 2 and daily thereafter. 2-ME administration in this protocol did not affect the retention of ^{131}I-A7 in the tumor on day 7 (Fig. 4B). Based on the results of these two experiments, we chose to use the second protocol for further RIT studies.

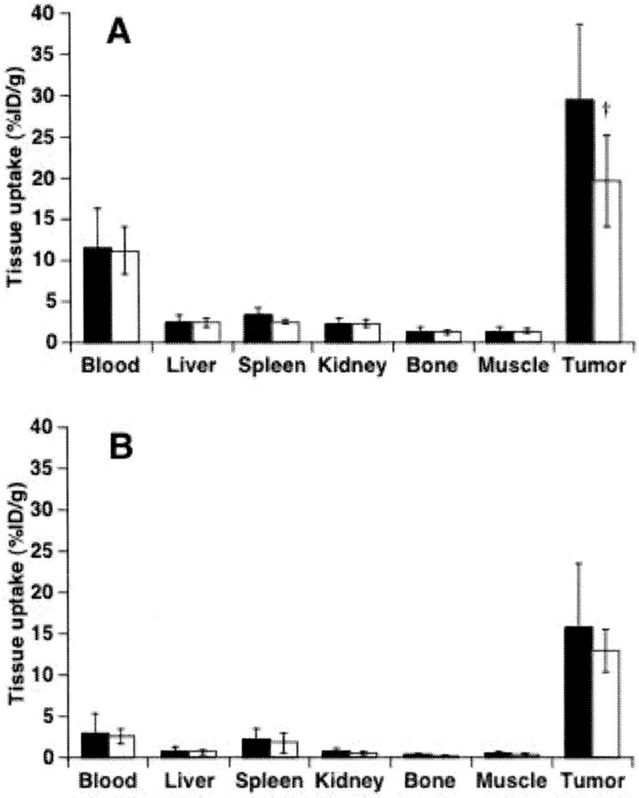

Fig. 4. Effect of 2-methoxyestradiol on biodistribution of ^{131}I-A7. A, on day 2 (n=10); B, on day 7 (n=5). B, Non-treated control; G, mice treated with 75 mg/kg/day p.o. of 2-methoxyestradiol. †, $P <$ 0.005.

2-ME treatment and RIT significantly suppressed the growth of xenografts (Fig. 5) in comparison to no treatment. The combined treatment of 2-ME and RIT demonstrated more efficient tumor growth inhibition than did monotherapy with 2-ME or RIT ($p < 0.05$). Tumor volumes on day 13, expressed as ratio relative to the initial volume, were 12.7 ± 2.95 in the control group, 4.73 ± 0.89 in the group administered 2-ME exclusively, 3.05 ± 0.37 in the group treated with RIT alone and 0.97 ± 0.20 in the group treated with RIT and 2-ME. Utility of 2-ME did not result in significantly different weight loss as compared to RIT alone (RIT alone, 11.3 ± 3.89%; RIT + 2-ME, 11.4 ± 3.90%). No animal showed signs of diarrhea or infection.

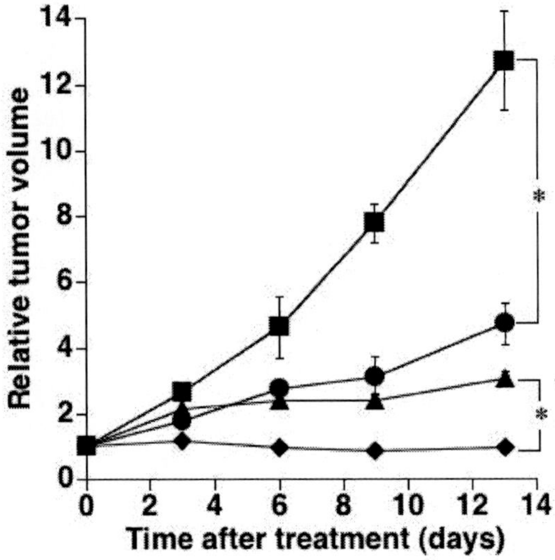

Fig. 5. Growth of LS180 colon cancer xenografts (n=10). Tumor volume is expressed as ratio to volume on day 0. B, Non-treated control; J, tumors treated with 75 mg/kg/day p.o. of 2-methoxyestradiol; H, tumors treated with 4.63 MBq of ^{131}I-A7; F, tumors treated with 2-methoxyestradiol and ^{131}I-A7. †, $P<0.05$.

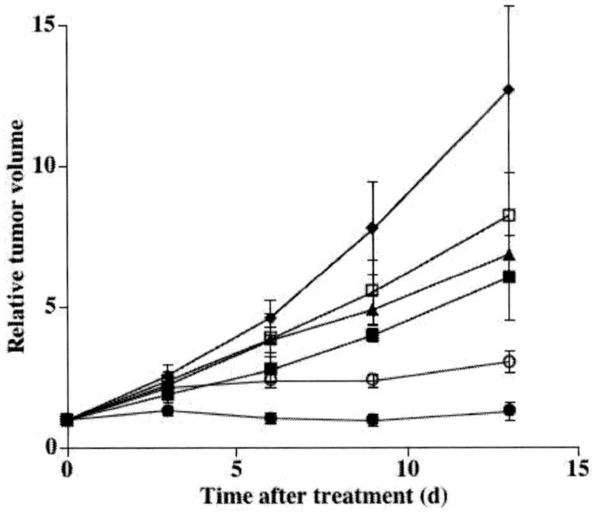

Fig. 6. Growth of LS180 colon cancer xenografts. Tumor volume is expressed as a ratio relative to the volume on day 0. F, Non-treated control; H, tumors treated with 200 mg/kg/day thalidomide; G, tumors treated with 4.63 MBq of ^{131}I-HPMS-1; B, tumors treated with thalidomide and ^{131}I-HPMS-1; E, tumors treated with 4.63 MBq of ^{131}I-A7; J, tumors treated with thalidomide and ^{131}I-A7.

Similarly, monotherapy of thalidomide treatment significantly suppressed growth of LS180 xenografts in comparison to no treatment ($p<0.005$) (Fig. 6). Relative tumor volumes on day 13 was 6.84 ± 0.71 in the group administered thalidomide. Treatment with a class-matched irrelevant MAb ^{131}I-HPMS-1 did not effectively suppress tumor growth. Combined treatment of thalidomide and ^{131}I-A7 RIT resulted in more efficient tumor growth inhibition than did monotherapy with thalidomide or RIT alone, displaying a tumor volume of 1.31 ± 0.33 on day 13 ($p < 0.005$ vs. RIT alone). Moreover, six of 10 tumors regressed with combined treatment, whereas no regression occurred with monotherapy.

Immunostaining of tumor sections with anti-factor VIII antibody demonstrated a reduction in microvessel density in tumors treated with 2-ME (7.9 ± 0.8/200× field) or thalidomide (8.30 ± 0.8) relative to control tumors (29.9 ± 2.5) ($P < 0.0001$). Interestingly, microvessel density in tumors treated with ^{131}I-A7 RIT (10.8 ± 1.4) was lower than that in control tumors ($p<0.0001$). Moreover, combination of RIT and thalidomide treatment further decreased microvessel density in tumors (4.3 ± 0.5; $p<0.0001$ vs. RIT monotherapy and $p<0.01$ vs. thalidomide monotherapy).

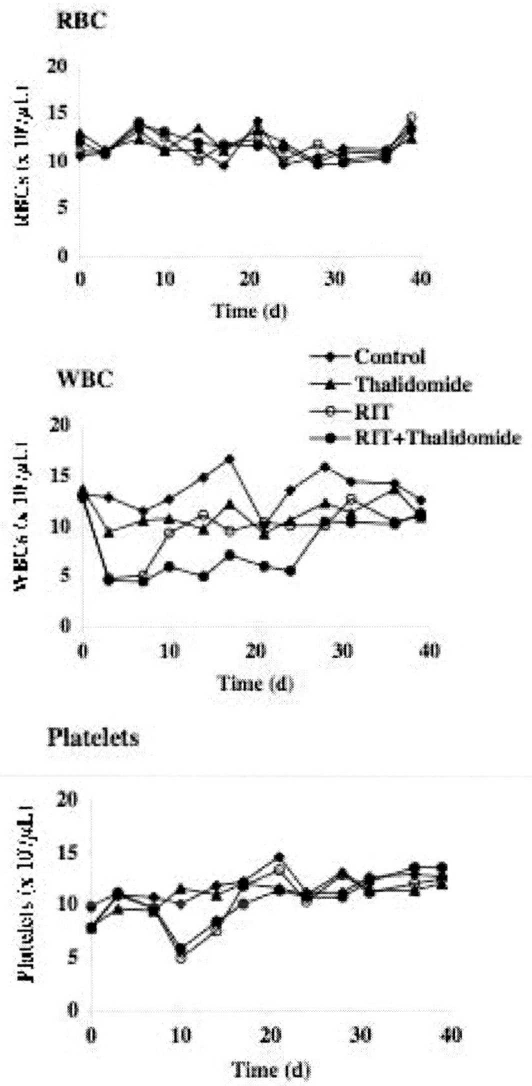

Fig. 7. Changes in peripheral blood cell counts in mice. (A), RBC; (B), WBC; (C), platelets. F, Non-treated control; H, tumors treated with 200 mg/kg/day thalidomide; E, tumors treated with 4.63 MBq RIT; J, tumors treated with thalidomide and RIT.

Thalidomide treatment induced minor depression of peripheral WBC count relative to that observed in non-treated mice (Fig. 7). The nadir of WBC count depression in mice treated with the combination of RIT and thalidomide therapy was similar to that in mice treated with RIT alone; however, the depression persisted in mice treated with the combination regimen. WBC count recovered following cessation of thalidomide administration. Changes in platelet count in mice receiving the combination treatment did not vary from those in mice undergoing RIT alone. Thalidomide administration did not affect RBC count. Use of thalidomide did not result in different weight loss as compared to the control (Fig. 8). Maximal body weight loss was similar in mice treated with RIT alone and those undergoing the combination regimen; however, recovery was delayed by thalidomide treatment. No mice showed signs of diarrhea or infection.

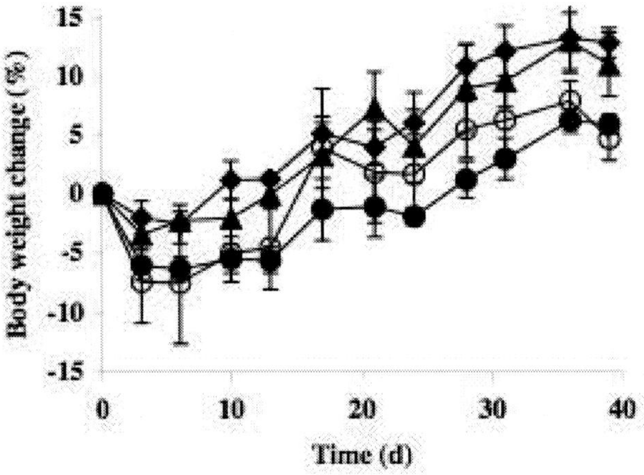

Fig. 8. Body weight changes of mice. F, Non-treated control; H, tumors treated with 200 mg/kg/day thalidomide; E, tumors treated with 4.63 MBq RIT; J, tumors treated with thalidomide and RIT.

In a Liver Metastasis Model

^{125}I-A7 accumulated almost exclusively in the LS180 liver metastases, demonstrating activity of 33.1 ± 19.8 %ID/g at two days and 6.8 ± 6.7 %ID/g at six days (Fig. 9). Accumulation of an irrelevant MAb, HMPS-1, in metastases was only 6.6 ± 4.2 %ID/g at two days; accumulation into normal organs was higher than that of A7 (Fig. 10). Consequently, target-to-nontarget ratios were approximately 7-10 times higher with ^{125}I-A7 (Table 2). Antiangiogenic therapy with 2-ME did not alter ^{125}I-A7 accumulation in metastases or normal tissues (Fig. 11). In Fig. 12, ^{125}I-A7 accumulation in liver metastases is plotted against total burden of metastases, which demonstrates the inverse correlation between A7 accumulation and metastasis burden in both non-treated and 2-ME-treated mice. No apparent difference in ^{125}I-A7 accumulation was observed between the two groups with regard to total weight of metastasis.

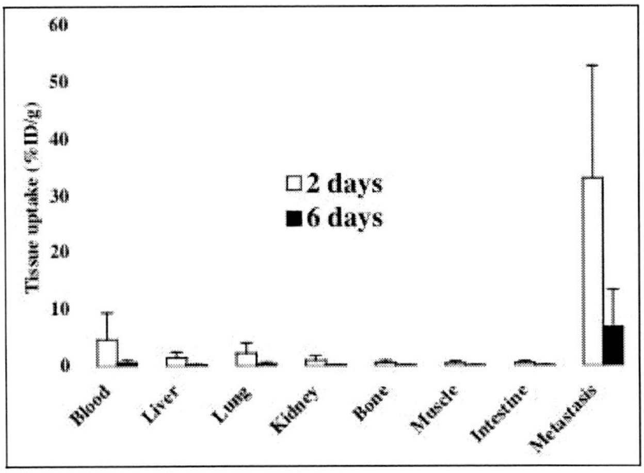

Fig. 9. Biodistribution of ^{125}I-A7 in mice bearing liver metastases (n=8-10).

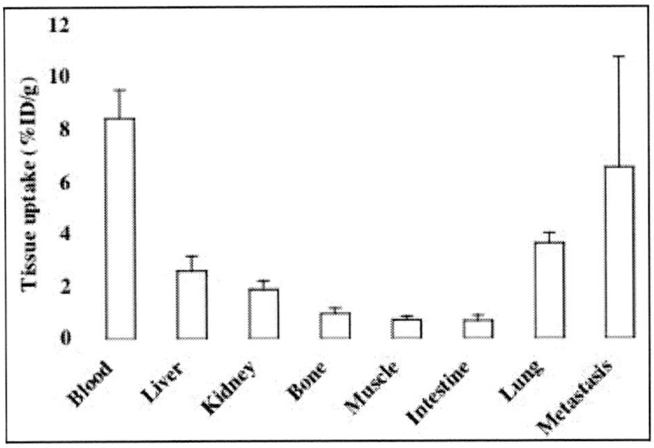

Fig. 10. Biodistribution of an irrelevant IgG1, ^{131}I-HPMS-1, in mice bearing liver metastases (n=4).

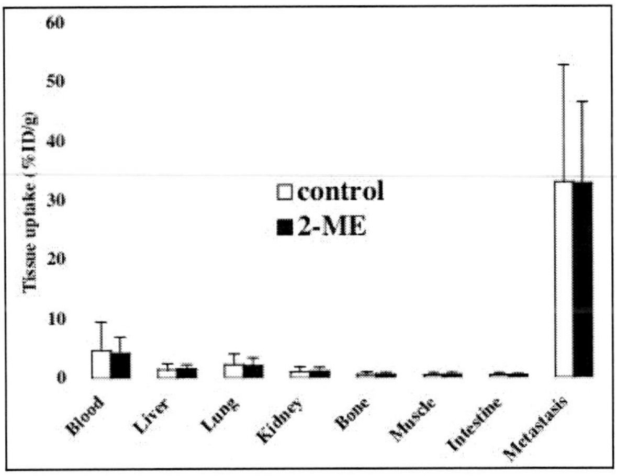

Fig. 11. Influence of 2-methoxyestradiol (2-ME) administration on ^{125}I-A7 accumulation in metastases. A dose of 75 mg/kg of 2-ME was administered one week after the cell inoculation and daily thereafter (n=5). ^{125}I-A7 was injected two weeks following the inoculation. Biodistribution was observed two days after ^{125}I-A7 injection.

Table 2. Tumor-to-Nontumor Ratio*.

	A7 (n=10)	HPMS-1 (n=4)
Blood	9.3 ± 3.2	0.8 ± 0.1
Liver	24.5 ± 8.7	2.6 ± 0.7
Kidney	35.8 ± 9.8	3.5 ± 0.5
Bone	61.3 ± 19.8	7.1 ± 1.8
Muscle	83.1 ± 25.5	9.2 ± 1.9
Lung	16.7 ± 5.4	1.8 ± 0.2
Intestine	72.1 ± 20.4	9.9 ± 1.8

*, Mean ± SD at 2 days after injection.

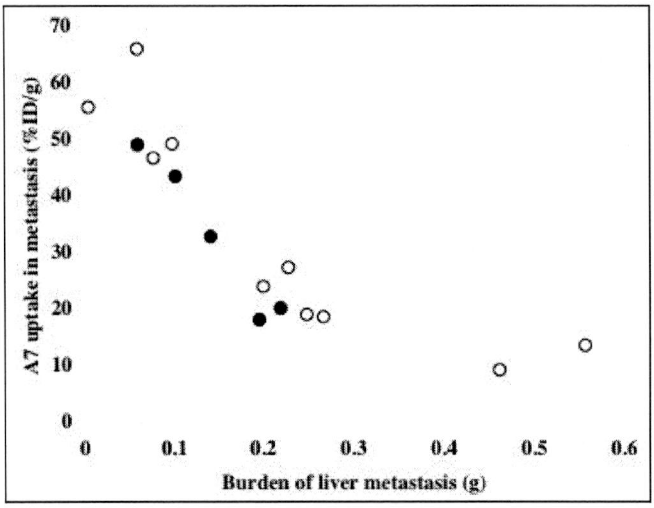

Fig. 12. Effect of metastatic burden on accumulation of ^{125}I-A7. Open circles, non-treated mice; closed circles, 2-ME-treated mice

Table 3. Therapeutic outcomes.

	n	Died*	Ascites†	Weight of metastases (g)‡
Non-treated control	6	1	5/5	5.96 ± 0.87
AT	8	0	4/8	2.67 ± 1.89
HPMS-1 RIT	6	2	3/4	2.25 ± 0.88
A7 RIT	8	0	1/8	0.85 ± 0.68
AT + HPMS-1 RIT	6	0	2/6	1.41 ± 0.68
AT + A7 RIT	8	0	0/8	0.28 ± 0.32

(Brackets in weight column: § Non-treated control vs AT; § Non-treated control vs HPMS-1 RIT; † HPMS-1 RIT vs A7 RIT; † AT + HPMS-1 RIT vs AT + A7 RIT; † A7 RIT vs AT + A7 RIT.)

*, Died before 4 weeks; †, number of mice with ascites at 4 weeks/number of surviving mice; ‡, weight at 4 weeks after the cell inoculation (mean ± SD of mice surviving for 4 weeks); §, $p<0.05$. AT, antiangiogenic therapy with 75 mg/kg/day 2-methoxyestradiol initiated at 1 week after the intrasplenic cell inoculation; HPMS-1 RIT, 7 MBq of an irrelevant ^{131}I-IgG at 2 weeks; A7 RIT, 7 MBq of a specific ^{131}I-IgG at 2 weeks.

Metastases occupied most of the liver in non-treated control mice four weeks following intrasplenic inoculation of LS180 cancer cells (Table 3). One mouse in the control group died due to metastases on day 26. Weight of metastases in the five surviving animals was 5.96 ± 0.87 g on day 28. Ascites due to metastases were observed in all control mice. 2-ME treatment inhibited the growth of metastases approximately 50% as compared with the control group. However, half of the animals in the antiangiogenic therapy group displayed ascites as a result of metastases at four weeks. Non-specific RIT with ^{131}I-HPMS-1 induced slight growth suppression; however, the effect was less significant than specific RIT with ^{131}I-A7. Moreover, two mice in the non-specific RIT group died on day 25, which was probably attributable to nonspecific radiation as well as metastases (Fig. 10 and Table 2). Three of four surviving mice in this group exhibited ascites. Combination of antiangiogenic therapy and specific RIT suppressed the growth of metastases more markedly than did monotherapy involving antiangiogenic therapy or RIT. Combination of AT and ^{131}I-HPMS-1 RIT was far less effective than the combination of antiangiogenic therapy with ^{131}I-A7 RIT.

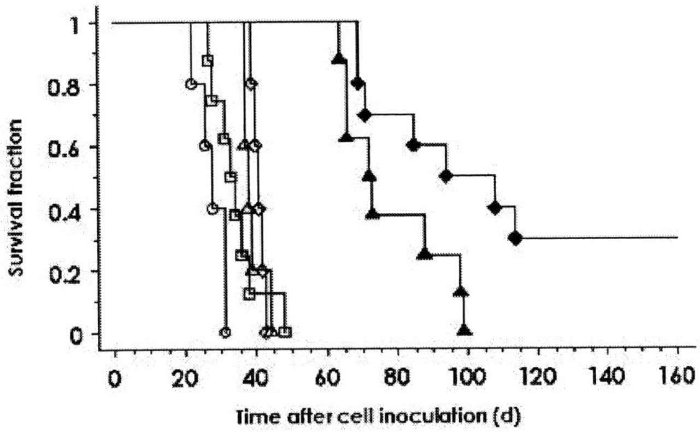

Fig. 13. Survival of mice bearing liver metastases of colon cancer cells. Circle, non-treated control; square, antiangiogenic therapy with daily 2-methoxyestradiol; open triangle, non-specific RIT with ^{131}I-HPMS-1; open diamond, combination of antiangiogenic therapy and non-specific RIT; closed triangle, specific RIT with ^{131}I-A7; closed diamond, combination of antiangiogenic therapy and specific RIT.

The investigation of survival of mice further confirmed the benefit of the combined therapy (Fig. 13). All non-treated mice died by 31 days after cell inoculation. Burden of liver metastases typically exceeded 5 g at autopsy. 2-ME antiangiogenic therapy alone resulted in slightly better survival of treated mice in comparison to that observed in untreated mice ($p<0.05$). Non-specific RIT with ^{131}I-HPMS-1 induced slight therapeutic effect ($p<0.01$); however, combination of antiangiogenic therapy and ^{131}I-HPMS-1 RIT failed to display appreciable benefit. ^{131}I-A7 RIT remarkably improved the survival of mice ($p<0.001$); however, all of these animals eventually died due to metastases. The combined regimen consisting of ^{131}I-A7 RIT and antiangiogenic therapy demonstrated superior therapeutic effect in comparison to monotherapy with either RIT or antiangiogenic treatment ($p<0.05$); three of 10 mice survived the entire observation period of 160 days. These findings are important because animal survival is a reflection of inhibitory effects on tumor growth as well as a toxicity profile of treatment that could be monitored only by an observation for a long period.

Antiangiogenic therapy involving daily administration of 2-ME did not induce appreciable myelotoxicity (Fig. 14). In addition, combination of antiangiogenic therapy with RIT apparently did not enhance toxicity of ^{131}I-A7 RIT.

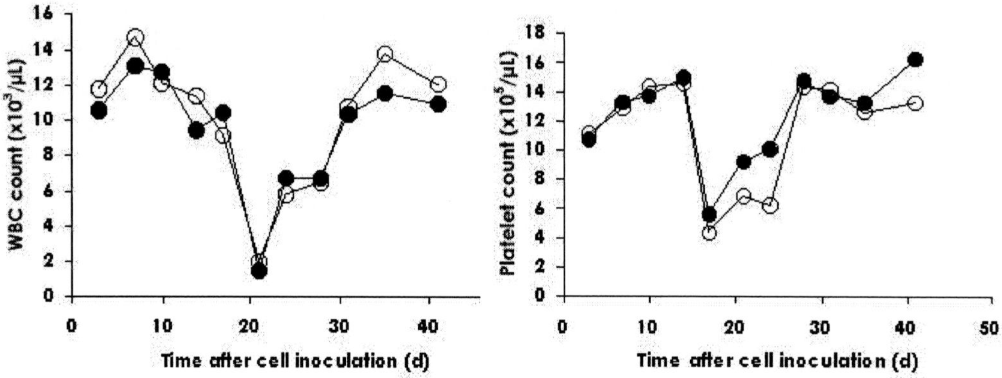

Fig. 14. Myelotoxicity of treatments. Open symbol, ^{131}I-A7 RIT; closed symbol, combination of RIT and antiangiogenic therapy.

Disucussion

Feasibility of combined use of RIT and antiangiogenic therapy to treat cancer lesions is verified by the series of animal experiments shown here. Typically, tumor growth beyond 1-3 mm in diameter leads to necrosis while further growth requires a new vascular supply from surrounding tissues [46]. Indeed, we confirmed the existence of microvessels within the xenografts and the antiangiogenic effect of 2-ME and thalidomide. These results indicate that the synergism between the cell killing by ß-irradiation with RIT and the inhibition of neoangiogenesis with 2-ME and thalidomide, produced better tumor growth suppression than monotherapy with RIT or antiangiogenic treatment. Furthermore, the direct anti-tumorigenic effect of 2-ME, probably the induction of apoptotic death of tumor cells, was added in the combination of RIT and 2-ME.

The results of the biodistribution analyses in subcutaneous xenografts suggest the importance of scheduling with administration of antiantiogenic agents. 2-ME administration initiated on the same day as the ^{131}I-A7 injection reduced ^{131}I-A7 accumulation within relatively large subcutaneous xenografts. In contrast, the retention of radioactivity within the tumor was not affected by the 2-ME manipulation. These results indicate that, to obtain the highest tumor radiation dose, antiangiogenic treatment may be required to be administered after the maximal tumor uptake is obtained. However, in a small liver metastasis model of the same cell line, the accumulation of ^{131}I-A7 in metastases was not affected by 2-ME administration initiated 2 days prior to the ^{131}I-A7 injection, suggesting that the accessibility of MAb to small metastatic lesions would be less influenced by antiangiogenic treatment than that to well-established tumors assessed. Therefore, in treating small metastatic lesions in adjuvant settings, the MAb injection would not need to precede administration of antiangiogenic agents, which certainly makes the administration schedule more flexible.

In theory, the advantages of combined therapy in the treatment of metastases include the following: a) RIT is more effective with respect to small tumors than large ones as small malignancies demonstrate greater and more homogeneous accumulation of labeled MAb than do large ones [17,24,47,48], b) antiangiogenic agents inhibit neovascularization that is required for the metastatic process [29], c) antiangiogenic agents and RIT act against different compartments of tumors, meaning that the former suppresses the endothelial compartment and indirectly inhibits the growth of tumor cells while the latter directly effects a killing of tumor cells, d) some of antiangiogenic agents also has the direct antitumorigenic effect, e) RIT may add damages to the vasculature in tumors, and f) both RIT and antiangiogenic treatment can be administered systemically. The metastatic ability of a tumor is dependent on angiogenesis at two stages of the metastatic cascade [29]. First, metastatic cells are not shed from a primary tumor until it has become neovascularized. Secondly, upon arrival at the target organ, metastatic emboli must undergo neovascularization if a metastasis is to grow to a clinically detectable size [29]. In the absence of neovascularization, metastatic foci remain limited in size. Acting together, antiangiogenic therapy may prolong the dormancy of microscopic metastases, whereas RIT may exterminate this population of cells.

The phenomena that ^{131}I-A7 RIT monotherapy reduced microvessel number in tumors as compared to no treatment, and, furthermore, the combined regimen more effectively decreased microvessel number than did thalidomide monotherapy were probably caused by radiation injury of vessels due to ^{131}I-A7. Because penetration of extravasated MAbs into solid tumors is disturbed by factors such as high interstitial pressure [49] and so-called binding site barrier [46]. MAbs would deposit close to the vasculature; for instance, IgG would require some days to cover the distance of ~1 mm after the extravasation [49]. Consequently, considerable radiation dose would be absorbed by the vascular compartment. In this sense, RIT and antiangiogenic therapy would cooperatively give effect to the suppression of tumor vasculature in the combined regimen.

There is a possible difficulty of RIT to treat microscopic metastases: the distribution of radiation dose within a tumor depends on tumor size and the energy of ionizing particles emitted by radionuclides [50,51]. In case of ß-emitters, mathematical model analyses estimate that a substantial portion of energy should deposits outside microscopic metastases because of millimeter path length of ß-particles, resulting in less favorable dosimetry. Nevertheless, several studies proved the effectiveness of RIT with ß-emitters in micrometastatic models [18-20,22]. Furthermore, it is suggested that radioantibodies are also effective on single avascular tumor cells [19]. These reports support the use of RIT in handling minimal disease. Radionuclides emitting Auger electrons or a-particles may be more favorable than ß-emitters in this aspect.

The effectiveness of RIT in small lesions, being contrary to the mathematical analyses, would be supported by following factors: firstly, MAb accumulation in tumors increases with decreasing tumor size [52,53], although mathematical models dealing with the size-effect on tumor curability with ß-emitters have assumed independence of tumor uptake from tumor size [50,51]. This phenomenon occurs not only in large tumors but also in lesions as small as those used in the current investigation [23,24,52,53]; indeed, we have previously demonstrated the size-dependency of MAb accumulation in the liver metastasis model of LS180 cells in the range of 0.1-100 mg [23]. Furthermore, homogeneity of MAb distribution would be improved in smaller nodules [17]. Secondly, better perfusion in smaller nodules and subsequent better oxygenation of tumor cells would improve response of cells to radiation [54]. In addition, satisfactory perfusion may allow cycling of cells, which may correlate with high radiation sensitivity [55]. Consequently, these favorable factors would compensate for the loss of b energy outside metastases, supporting the use of ß emitter in an adjuvant setting.

The daily dose of 2-ME (75 mg/kg) did not contribute appreciable toxicity to mice treated with RIT assessed by body weight change. This result was also shown in different strains of C3H mice [56] and SCID mice [39]. Bone marrow toxicity, which is a major complication of RIT, has not been reported with this agent. In addition, as a nontoxic estrogen metabolite occurring in normal human urine, 2-ME has virtually no effect on non-proliferating cells. These facts indicate that the combined utility of 2-ME and RIT is attractive not only in terms of efficacy, but also in terms of toxicity profile. Furthermore, 2-ME is effective even upon oral dosing, which simplifies its administration. In contrast, thalidomide displays several adverse effects, including somnolence, constipation, skin rashes, orthostatic hypotension, nausea, peripheral neuropathy and neutropenia in addition to teratogenicity [40,57]. The majority of these untoward effects do not occur in RIT therapy; however, neutropenia may affect combination with RIT. Minor depression of peripheral WBC count was observed in mice treated with daily thalidomide. The combined regimen delayed recovery of WBC count in comparison to RIT monotherapy. Recovery of body weight loss was also delayed by this combination. Reports describing clinical trials of thalidomide indicate that thalidomide-induced neutropenia rarely occurs in patients other than those presenting with human immunodeficiency virus infection and recipients of allogeneic bone marrow transplant [40,57]; however, marrow toxicity should be carefully monitored during combination therapy involving RIT and thalidomide.

In conclusion, we demonstrated the synergistic enhancement of therapeutic efficacy with RIT and antiangiogenic treatment in a subcutaneous xenograft model and a small liver metastasis model of human colon cancer cells. These results indicate the feasibility of this novel combination treatment consisting of RIT and antiangiogenic therapy initiated at the early stage of metastasis.

References

[1] Kohler G, Milstein C. Continuous cultures of fused cells secreting antibody of predefined specificity. *Nature* 1975;256:495-497.

[2] Oriuchi N, Higuchi T, Hanaoka H, Iida Y, Endo K. Current status of cancer therapy with radiolabeled monoclonal antibody. *Ann Nucl Med* 2005;19:355-365.

[3] Reilly RM. Radioimmunotherapy of solid tumors: the promise of pretargeting strategies using bispecific antibodies and radiolabeled haptens. *J Nucl Med* 2006;47:196-199.

[4] Koppe MJ, Postema EJ, Aarts F, Oyen WJ, Bleichrodt RP, Boerman OC. Antibody-guided radiation therapy of cancer. *Cancer Metastasis Rev* 2005;24:539-567.

[5] Goldenberg DM. Targeted therapy of cancer with radiolabeled antibodies. *J Nucl Med* 2002;43:693-713.

[6] Rajendran J, Gopal A, Durack L, Fisher D, Press O, Eary J. Comparison of radiation dose estimation for myeloablative radioimmunotherapy for relapsed or recurrent mantle cell lymphoma using ^{131}I tositumomab to that of other types of non-Hodgkin's lymphoma. *Cancer Biother Radiopharm* 2004;19:738-745.

[7] Kaminski MS, Zasadny KR, Francis IR, et al. Iodine-131-anti-B1 radioimmunotherapy for B-cell lymphoma. *J Clin Oncol* 1996;14:1974-1981.

[8] Kelly MP, Lee FT, Smyth FE, Brechbiel MW, Scott AM. Enhanced efficacy of ^{90}Y-radiolabeled anti-Lewis Y humanized monoclonal antibody hu3S193 and paclitaxel combined-modality radioimmunotherapy in a breast cancer model. *J Nucl Med* 2006;47:716-725.

[9] Kinuya S, Yokoyama K, Tega H, et al. Efficacy, toxicity and mode of interaction of combination radioimmunotherapy with 5-fluorouracil in colon cancer xenografts. *J Cancer Res Clin Oncol* 1999;125:630-636.

[10] Masters GR, Berger MA, Albone EF. Synergistic effects of combined therapy using paclitaxel and [^{90}Y-DOTA]776.1 on growth of OVCAR-3 ovarian carcinoma xenografts. *Gynecol Oncol* 2006.

[11] Kinuya S, Yokoyama K, Hiramatsu T, et al. Combination radioimmunotherapy with local hyperthermia: increased delivery of radioimmunoconjugate by vascular effect and its retention by increased antigen expression in colon cancer xenografts. *Cancer Lett* 1999;140:209-218.

[12] Mittal BB, Zimmer MA, Sathiaseelan V, et al. Phase I/II trial of combined ^{131}I anti-CEA monoclonal antibody and hyperthermia in patients with advanced colorectal adenocarcinoma. *Cancer* 1996;78:1861-1870.

[13] Kinuya S, Yokoyama K, Kawashima A, et al. Pharmacologic intervention with angiotensin II and kininase inhibitor enhanced efficacy of radioimmunotherapy in human colon cancer xenografts. *J Nucl Med* 2000;41:1244-1249.

[14] Nakamura K, Kubo A. Effect of interleukin-2 on the biodistribution of technetium-99m-labelled anti-CEA monoclonal antibody in mice bearing human tumour xenografts. *Eur J Nucl Med* 1994;21:924-929.

[15] Kuhn JA, Beatty BG, Wong JY, et al. Interferon enhancement of radioimmunotherapy for colon carcinoma. *Cancer Res* 1991;51:2335-2339.

[16] Saga T, Sakahara H, Nakamoto Y, et al. Radioimmunotherapy for liver micrometastases in mice: pharmacokinetics, dose estimation, and long-term effect. *Jpn J Cancer Res* 1999;90:342-348.

[17] Sato N, Saga T, Sakahara H, et al. Intratumoral distribution of radiolabeled antibody and radioimmunotherapy in experimental liver metastases model of nude mouse. *J Nucl Med* 1999;40:685-692.

[18] Sharkey RM, Weadock KS, Natale A, et al. Successful radioimmunotherapy for lung metastasis of human colonic cancer in nude mice. *J Natl Cancer Inst* 1991;83:627-632.

[19] Blumenthal RD, Sharkey RM, Haywood L, et al. Targeted therapy of athymic mice bearing GW-39 human colonic cancer micrometastases with ^{131}I-labeled monoclonal antibodies. *Cancer Res* 1992;52:6036-6044.

[20] Dunn RM, Juweid M, Sharkey RM, Behr TM, Goldenberg DM. Can occult metastases be treated by radioimmunotherapy? *Cancer* 1997;80:2656-2659.

[21] Vogel CA, Galmiche MC, Buchegger F. Radioimmunotherapy and fractionated radiotherapy of human colon cancer liver metastases in nude mice. *Cancer Res* 1997;57:447-453.

[22] Behr TM, Blumenthal RD, Memtsoudis S, et al. Cure of metastatic human colonic cancer in mice with radiolabeled monoclonal antibody fragments. *Clin Cancer Res* 2000;6:4900-4907.

[23] Kinuya S, Yokoyama K, Kawashima A, et al. Radioimmunotherapy with ^{186}Re-labeled monoclonal antibody to treat liver metastases of colon cancer cells in nude mice. *Cancer Biother Radiopharm* 2002;17:681-687.

[24] Li XF, Kinuya S, Yokoyama K, et al. Benefits of combined radioimmunotherapy and anti-angiogenic therapy in a liver metastasis model of human colon cancer cells. *Eur J Nucl Med Mol Imaging* 2002;29:1669-1674.

[25] Kinuya S, Li XF, Yokoyama K, et al. Intraperitoneal radioimmunotherapy in treating peritoneal carcinomatosis of colon cancer in mice compared with systemic radioimmunotherapy. *Cancer Sci* 2003;94:650-654.

[26] Kinuya S, Yokoyama K, Izumo M, et al. Feasibility of ^{186}Re-radioimmunotherapy for treatment in an adjuvant setting of colon cancer. *J Cancer Res Clin Oncol* 2003;129:392-396.

[27] Kinuya S, Yokoyama K, Izumo M, et al. Locoreginal radioimmunotherapy with ^{186}Re-labeled monoclonal antibody in treating small peritoneal carcinomatosis of colon cancer in mice in comparison with ^{131}I-counterpart. *Cancer Lett* 2005;219:41-48.

[28] Hird V, Maraveyas A, Snook D, et al. Adjuvant therapy of ovarian cancer with radioactive monoclonal antibody. *Br J Cancer* 1993;68:403-406.

[29] Folkman J. Angiogenesis in cancer, vascular, rheumatoid and other disease. *Nat Med* 1995;1:27-31.

[30] Folkman J. Antiangiogenesis in cancer therapy--endostatin and its mechanisms of action. *Exp Cell Res* 2006;312:594-607.

[31] Duda DG. Antiangiogenesis and drug delivery to tumors: bench to bedside and back. *Cancer Res* 2006;66:3967-3970.

[32] Mancuso A, Sternberg CN. Colorectal cancer and antiangiogenic therapy: what can be expected in clinical practice? *Crit Rev Oncol Hematol* 2005;55:67-81.

[33] Kinuya S, Kawashima A, Yokoyama K, et al. Cooperative effect of radioimmunotherapy and antiangiogenic therapy with thalidomide in human cancer xenografts. *J Nucl Med* 2002;43:1084-1089.

[34] Kinuya S, Kawashima A, Yokoyama K, et al. Anti-angiogenic therapy and radioimmunotherapy in colon cancer xenografts. *Eur J Nucl Med* 2001;28:1306-1312.

[35] Kinuya S, Yokoyama K, Koshida K, et al. Improved survival of mice bearing liver metastases of colon cancer cells treated with a combination of radioimmunotherapy and antiangiogenic therapy. *Eur J Nucl Med Mol Imaging* 2004;31:981-985.

[36] Kotanagi H, Takahashi T, Masuko T, Hashimoto Y, Koyama K. A monoclonal antibody against human colon cancers. *Tohoku J Exp Med* 1986;148:353-360.

[37] Hirano K, Iiizumi Y, Hayashi Y, et al. A highly sensitive assay method for human placental alkaline phosphatase involving a monoclonal antibody bound to a paper disk. *Anal Biochem* 1986;154:624-631.
[38] Asano M, Yukita A, Suzuki H. Wide spectrum of antitumor activity of a neutralizing monoclonal antibody to human vascular endothelial growth factor. *Jpn J Cancer Res* 1999;90:93-100.
[39] Klauber N, Parangi S, Flynn E, Hamel E, D'Amato RJ. Inhibition of angiogenesis and breast cancer in mice by the microtubule inhibitors 2-methoxyestradiol and taxol. *Cancer Res* 1997;57:81-86.
[40] D'Amato RJ, Loughnan MS, Flynn E, Folkman J. Thalidomide is an inhibitor of angiogenesis. *Proc Natl Acad Sci U S A* 1994;91:4082-4085.
[41] Kenyon BM, Browne F, D'Amato RJ. Effects of thalidomide and related metabolites in a mouse corneal model of neovascularization. *Exp Eye Res* 1997;64:971-978.
[42] Kruse FE, Joussen AM, Rohrschneider K, Becker MD, Volcker HE. Thalidomide inhibits corneal angiogenesis induced by vascular endothelial growth factor. *Graefes Arch Clin Exp Ophthalmol* 1998;236:461-466.
[43] Reiser F, Way D, Bernas M, Witte M, Witte C. Inhibition of normal and experimental angiotumor endothelial cell proliferation and cell cycle progression by 2-methoxyestradiol. *Proc Soc Exp Biol Med* 1998;219:211-216.
[44] Amorino GP, Freeman ML, Choy H. Enhancement of radiation effects in vitro by the estrogen metabolite 2-methoxyestradiol. *Radiat Res* 2000;153:384-391.
[45] Weidner N, Semple JP, Welch WR, Folkman J. Tumor angiogenesis and metastasis--correlation in invasive breast carcinoma. *N Engl J Med* 1991;324:1-8.
[46] Weinstein JN, van Osdol W. Early intervention in cancer using monoclonal antibodies and other biological ligands: micropharmacology and the "binding site barrier". *Cancer Res* 1992;52:2747s-2751s.
[47] Juweid ME, Sharkey RM, Behr T, et al. Radioimmunotherapy of patients with small-volume tumors using iodine-131-labeled anti-CEA monoclonal antibody NP-4 F(ab')2. *J Nucl Med* 1996;37:1504-1510.
[48] Gold DV, Cardillo T, Vardi Y, Blumenthal R. Radioimmunotherapy of experimental pancreatic cancer with ^{131}I-labeled monoclonal antibody PAM4. *Int J Cancer* 1997;71:660-667.
[49] Jain RK. Delivery of novel therapeutic agents in tumors: physiological barriers and strategies. *J Natl Cancer Inst* 1989;81:570-576.
[50] O'Donoghue JA, Bardies M, Wheldon TE. Relationships between tumor size and curability for uniformly targeted therapy with beta-emitting radionuclides. *J Nucl Med* 1995;36:1902-1909.
[51] Fujimori K, Furudate M. Dosimetry and radiation dose distribution in tumors for radioimmunotherapy: the effect of tumor size. *Kaku Igaku* 1994;31:241-248.
[52] Hagan PL, Halpern SE, Dillman RO, et al. Tumor size: effect on monoclonal antibody uptake in tumor models. *J Nucl Med* 1986;27:422-427.
[53] Behr TM, Sharkey RM, Juweid ME, et al. Variables influencing tumor dosimetry in radioimmunotherapy of CEA-expressing cancers with anti-CEA and antimucin monoclonal antibodies. *J Nucl Med* 1997;38:409-418.

[54] Okunieff P, Dols S, Lee J, et al. Angiogenesis determines blood flow, metabolism, growth rate, and ATPase kinetics of tumors growing in an irradiated bed: ^{31}P and ^{2}H nuclear magnetic resonance studies. *Cancer Res* 1991;51:3289-3295.

[55] Hall E. *Radiobiology for radiologist.* 5th ed. Philadelphia: Lippincott Williams & Wilkins; 2000.

[56] Fotsis T, Zhang Y, Pepper MS, et al. The endogenous oestrogen metabolite 2-methoxyoestradiol inhibits angiogenesis and suppresses tumour growth. *Nature* 1994;368:237-239.

[57] Calabrese L, Fleischer AB. Thalidomide: current and potential clinical applications. *Am J Med* 2000;108:487-495.

In: New Research on Angiogenesis Inhibitors
Editor: Elmer T. Skinard, pp. 99-113
ISBN: 1-60021-473-8

Chapter V

QUINONES AS ANTIANGIOGENIC AGENTS

Manal Chatila **and Hala Gali-Muhtasib***
Department of Biology, American University of Beirut, Beirut, Lebanon

ABSTRACT

Quinones are widely distributed among all respiring organisms. They participate in important reactions including photosynthesis, respiration and blood coagulation. Quinones have been used as antimalarial, antibacterial and anticancer agents. In addition to their redox-cycling capacity, they have been found to influence various signaling pathways known to affect cell proliferation, apoptosis and migration. Some quinones have been shown to decrease the angiogenic response, which is a prerequisite for tumor growth and metastasis. In this review, we attempt to describe the antitumor and antiangiogenic potential of benzoquinone ansamysins, anthracyclines, naphthoquinones, in addition to other interesting quinones.

1. INTRODUCTION

1.1. Quinones Reactivity and Redox-cycling

Quinones are generally defined as molecules containing a benzene moiety with a double ketone functional group. By virtue of the quinone moiety, these agents can undergo a biochemical reduction by one or two electrons, and this is catalyzed by flavoenzymes in the organism using NADPH as an electron donor. This bioreductive process leads to the semiquinone and subsequent reactions with oxygen, all of which are believed to be responsible for most of the drug's activity. The process of reductive activation is the first step

* Correspondence concerning this article should be addressed to Department of Biology, American University of Beirut, Beirut, Lebanon. Fax: +9611 744461, E-mail: amro@aub.edu.lb.

in the process of redox cycling because bioreduction leads to the semiquinone of the anticancer agent which, under aerobic conditions, is oxidized by molecular oxygen to the parent compound, a process that results in the concomitant production of superoxide radical anion ($O_2^{\cdot -}$), and continues until the system becomes anaerobic.

The formation of $O_2^{\cdot -}$ is the beginning of a cascade that generates H_2O_2 and hydroxyl radicals ($^{\cdot}OH$), generally referred to as reactive oxygen species (ROS). Superoxide radical anions are unstable in aqueous solution and dismutate to hydrogen peroxide (H_2O_2) and molecular oxygen. Hydrogen peroxide reacts with iron (Fenton reaction), and generates extremely toxic hydroxyl radicals. Most of the cytotoxic effects caused by ROS can be attributed to these hydroxyl radicals. Cells continuously generate ROS during normal aerobic metabolism, but they are equipped with antioxidant enzymes such as superoxide dismutase, catalase or glutathione peroxidase and radical scavengers such as ascorbic acid or glutathione; it is the excessive release of ROS that leads to a prooxidant balance that accounts for the pathological effects produced by these species (Asche, 2005, O'Brien, 1991). ROS can oxidize cellular macromolecules, and are thus responsible for the damage occurring during oxidative stress.

In addition to redox-cycling, quinones react with cellular nucleophiles in a Michael-type 1,4-addition. In particular, they react with glutathione, which is an important player in the cell's antioxidant defense system. Exposing the cells to high levels of quinones may lead to the depletion of GSH, and thus to the alkylation of proteins by quinones, causing irreversible damage and cell death (Asche, 2005; O'Brien, 1991).

1.2. Quinones Multiple Biological Effects

Quinones constitute an important class of naturally occurring and synthetic compounds with an enormous variety of functions. Quinones such as Plastoquinone serve as electron acceptors in electron transport chains in Photosystems I & II of photosynthesis, and are thus widely distributed among plants. Coenzyme Q, or Ubiquinone functions as an electron carrier in the respiratory chain. It is extensively found, as its name implies, "everywhere" in respiring organisms. Vitamin K, a naphthoquinone derivative, is required for blood coagulation and participates in the carboxylation of glutamate to γ-carboxyglutamate. In addition, quinones such as the naphthoquinones juglone from *Juglans nigra* and plumbagin from *Plumbago rosea*, exhibit growth inhibitory effects on bacteria or fungi and are used by plants as defensive compounds. Animals also use quinones as a defense mechanism; the bombardier beetle shoots a hot cloud of quinone towards an oncoming enemy (Asche, 2005; O'Brien, 1991). In addition to being used as antimalarial and antibacterial agents, and as fungicides, quinones comprise one of the largest classes of antitumor agents (O'Brien, 1991). For instance, the anthracycline antibiotics are among the most utilized anticancer agents ever developed (Asche, 2005).

Besides being antineoplastic agents, different classes of quinones are being studied as angiogenesis inhibitors (Table 1). Angiogenesis is the formation of new blood vessels from preexisting ones; it is crucial for tumor growth and metastasis. Tumors must continuously stimulate the growth of new capillary blood vessels for them to grow. These new blood vessels may as well provide a gateway for tumor cells to enter the circulation and to metastasize to distant sites (Folkman and Shing, 1992). In response to low oxygen levels, the levels of hypoxia-inducible factor-1α (HIF-1α) increase; HIF-1α is a transcription factor that induces the expression of more than 40 proteins (Semenza, 2002), the most important of which is vascular endothelial growth factor (VEGF). VEGF binds to its receptors on endothelial cells; it causes them to proliferate and be recruited to the avascularized areas within the tumor where they form capillaries, and thereby ensure proper vasculature and food supply for the tumor to grow, and later to metastasize.

Table 1. Molecular targets of anti-angiogenic quinones.

Quinone	Family	Molecular targets and distinctive characteristics	Effects on Angiogenesis	References
Geldanamycin	Benzoquinone Ansamycins (BA)	Binds N-terminal domain of Hsp-90 molecular chaperone	Inhibits FAK phosphorylation, lowers HIF-1α protein and mRNA levels, affects oncogenic signaling proteins	Whitesell *et al*, 1994; Zagzag *et al.*, 2003; Neckers and Neckers, 2002; Mabjeesh *et al.*, 2002
Allylamino-17-demethoxygeld-anamycin	BA	Same; lower *in vivo* toxicity, increased stability; clinical trials	Same, also reduces secretion of uPA	Sanderson *et al.*, 2006
Doxorubicin	Anthracycline		Targets tumor vasculature when administered in an NGR-coupled liposome, decreases VEGF-mediated angiogenesis	Pastorino *et al.*, 2003; Lennernas *et al.*, 2003
Vitamin K2	Naphthoquinone		Suppresses endothelial cell proliferation, supresses neovascularization when combined with Perindopril	Yoshiji *et al.*, 2005
Shikonin	Naphthoquinone		Decreases TNF-α-and tumor-induced angiogenesis	Hisa *et al.*, 1998
Thymoquinone			Decreases levels of HIF-1α, VEGF secretion, cancer cell migration	Gali-Muhtasib *et al.*, 2006
Cochlioquinone A1			Inhibits FGF-induced invasion of bovine aortic endothelial cells and tube formation on Matrigel	Jung *et al.*, 2003
Cannabidiol Hydroxyquinone	Cannabinoids		Causes apoptosis of vascular endothelial cells and inhibits angiogenesis in *ex vivo* and *in vivo* models	Kogan *et al.*, 2004
Linomide		Inhibits TNF-α secretion by tumor-associated macrophages	Inhibits tumor angiogenesis in a variety of rat and human prostatic cancers independent of growth rate, androgen sensitivity, or metastatic ability	Vukanovic and Isaacs, 1995; Gourley and Williamson, 2000

It would thus be of clinical importance if antineoplastic quinones were to possess antiangiogenic properties as well. In this review, we describe three classes of quinones: Benzoquinone Ansamycins, Anthracyclines and Naphthoquinones, in addition to their mechanism of action as anticancer agents. We also focus on some studies of selected quinones with antiangiogenic potential.

2. Antineoplastic and Antiangiogenic Activities of Selected Quinones

2.1. Benzoquinone Ansamycins

Benzoquinone Ansamycins (BA) are a family of antibiotics characterized by linkage of a quinone moiety to a planar macrocyclic ansa bridge structure. Typical BA are Geldanamycin and Herbimycin, which were first isolated in the 1970s (Figure 1). These were first shown to reverse the transformation by v-src (Uehara *et al.*, 1986). They were also shown to possess potent antitumor activity against human cancer cells both *in vitro* and in xenograft models (Whitesell *et al.,* 1992). Their mode of action was originally attributed to direct inhibition of Tyrosine Kinase, but it wasn't until 1994 that Whitesell *et al.* showed that Heat Shock Protein 90 (Hsp90) is a primary target of these quinones. Hsp90 is a molecular chaperone critical for the folding, stability and function of many proteins including oncogenic signaling proteins that promote the growth and the survival of cancer cells. Hsp90 clients include mutant p53, Her-2/ErbB2, c-Raf-1, v-Src, Bcr/Abl, c-Met and Cdk4 (Asche, 2005; Neckers and Neckers, 2002). These antibiotics bind to a highly conserved 25 kDa N-terminal domain of Hsp90, a binding site for ATP, resulting in disruption of the Hsp90 chaperone complex. The inhibition of this complex blocks multiple oncogenic pathways leading to the depletion and degradation of a subset of proteins involved in the progression of cancer (Goetz *et al.*, 2003). Both these antibiotics did not make it to clinical trials because of their *in vivo* toxicity and instability; instead, 17 Allylamino-17-demethoxygeldanamycin (17-AAG), and the soluble derivative 17-(dimethylaminoethylamino)-17-demethoxygeldanamycin (17-DMAG) possess lower in vivo toxicity and increased stability in comparison with GA. Both compounds have entered phase I clinical trials (Sanderson *et al.,* 2006).

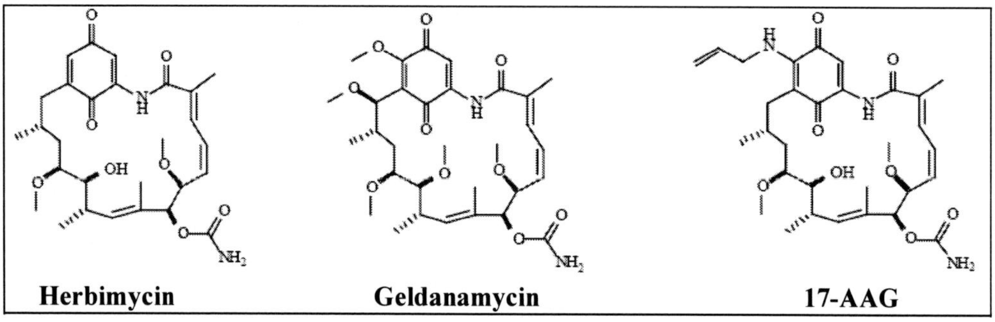

Figure 1. Chemical structures of selected Benzoquinone Ansamycins

Given the fact that some of Hsp90 client proteins are involved in glioma invasion, Zagzag *et al.* have tested for the potential of BAs to be used for treating invasive glioblastomas. They found that GA decreased migration of three different glioma cell lines at non-toxic levels, regardless of p53 status (Zagzag *et al.*, 2003). As many gliomas have alterations in these genes, this makes GA an important candidate for glioma treatment (Kleihues and Cavenee, 2000). This effect was associated with BA's ability to inhibit Hsp90. In addition, it was shown that GA inhibits focal adhesion kinase (FAK) phosphorylation. FAK is the initial protein to be tyrosine-phosphorylated in the integrin signaling pathway, and it plays an important role in cell migration and angiogenesis (Burridge *et al.*, 1992). This effect was also associated with GA's ability to inhibit Hsp90, since overexpression of this chaperone was able to reverse the inhibitory effect that GA had on FAK phosphorylation and migration (Rousseau *et al.*, 2000). The authors have also shown that GA caused degradation of Cobalt Chloride-induced HIF-1α proteins; HIF-1α being a transcription factor that regulates expression of the potent angiogenic cytokine VEGF, GA's ability to decrease the levels of this protein was associated with its ability to inhibit cell migration (Zagzag *et al.*, 2003).

In support of this last notion, Mabjeesh *et al.*, have shown that GA reduces nuclear HIF-1α protein levels in prostate and other common cancer cells in a time- and dose-dependent manner under both oxic and hypoxic conditions. This degradation is mediated by the proteosome pathway, as the use of proteosome-selective inhibitor LCN prevented HIF-1α degradation (Mabjeesh *et al.*, 2002). In addition, GA was found to inhibit HIF-1α transcriptional activity, and that was independent of GA's ability to induce its degradation; blocking GA-induced degradation with LCN failed to restore HIF-1 transcriptional activity (Mabjeesh *et al.*, 2002).

In another study by Sanderson *et al.*, GA and its derivatives were tested. All three compounds inhibited production of VEGF-A by tumor cells, and blocked the proliferative response of human endothelial cells at nanomolar levels, while sparing the quiescent vasculature cells (Sanderson *et al.*, 2006). 17-AAG was found to inhibit endothelial cell motility and migration in a concentration-dependent manner, without affecting cell viability; when VEGF was used as a chemoattractant in migration and motility assays, the extent of inhibition was greater, suggesting that Hsp90 exerts major effects on endothelial cell migration/motility via VEGF receptors (Sanderson *et al.*, 2006). Again, FAK phosphorylation was inhibited by 17-AAG, in addition to other Hsp90 client proteins that affect motility; both GA and 17-AAG down-regulated AKT, which activates Cdc42 and Rac1, in order to affect cell motility (Sanderson *et al.*, 2006). In addition to inhibiting motility, 17-AAG was found to reduce the secretion of urokinase-type plasminogen activator (uPA) in a concentration-dependent manner; uPA is a protease that cleaves plasminogen to plasmin upon binding to its receptor, causing degradation of the extracellular matrix by activating matrix metalloproteinases (Sanderson *et al.*, 2006). The authors of the paper also show that BA treatment leads to VEGF receptor-2 downregulation in endothelial cells and blood vessels within human tumor xenograft models; this receptor plays an important role in mitogenesis, migration, invasion and differentiation of endothelial cells, and was found to coassociate with Hsp90 (Sanderson *et al.*, 2006).

2.2. Anthracyclines

Anthracyclines rank among the most effective anticancer drugs ever developed (Weiss, 1992). The first anthracyclines were isolated from the pigment-producing *Streptomyces peucetius* in the 1960s and were named doxorubicin (DOX) and daunorubicin (DNR) (Figure 2). Whereas DOX is an essential component of treatment of breast cancer, childhood solid tumors, soft tissue sarcomas, and aggressive lymphomas, DNR shows activity in acute lymphoblastic or myeloblastic leukemias. As with any other anticancer agent, however, the clinical use of both DOX and DNR soon proved to be hampered by serious problems such as the development of resistance in tumor cells or toxicity in healthy tissues (Minotti *et al.,* 2004). Delivering anthracyclines as prodrugs to tumor tissue may help diminish toxicity; so far, promising results have been obtained with a peptidic derivative of DOX (Fernandez *et al.,* 2001). Anthracyclines share three principal functional components important for biological activity: the aglycone intercalator (tetracyclic ring with quinone moiety), ring A with its anchoring function and the amino sugar. Although intercalation is a necessary requirement for binding to DNA and even may contribute to their anticancer properties, it is not sufficient on its own. The aglycone itself does not exhibit biological activity (Asche, 2005).

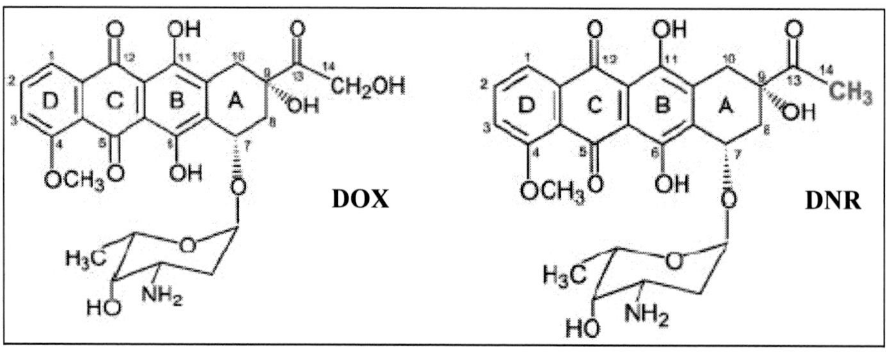

Figure 2. Structures of the Anthracyclines, Doxorubicin and Daunorubicin

There is now consensus that the antitumor activity of these anthracyclines mainly results from an inhibition of the mammalian topoisomerase II. Topoisomerases are nuclear enzymes regulating DNA topology through single- (topoisomerase I) or double-strand breaks (topoisomerase II). They are extensively involved in DNA metabolism including replication and transcription. Anthracyclines convert topo II into an endogenous toxin by stabilising the topo II-bridged DNA strand breaks in which the enzyme is covalently linked to the 5'-end of the cleaved phosphodiester bond of DNA ("cleavable complex"). They produce persistent cleavable complexes leading to an extensive increase in DNA strand breaks, which stimulate various cellular processes including apoptotis (Asche, 2005).

Pastorino *et al.* have exploited the tumor microenvironment to target the anthracycline DOX to tumor cells and their vasculature; unique epitopes expressed on tumor endothelial cells can function as targets for the anti-angiogenic therapy of the tumor. An NGR peptide that targets aminopeptidase N, a marker of angiogenic endothelial cells, was coupled to the surface of liposomal DOX, and was used to treat orthotopic neuroblastoma (NB) xenografts in SCID mice. Histopathological analysis of cryosections taken from treated mice revealed pronounced destruction of the tumor vasculature with a marked decreased in vessel density. Double staining of tumors with terminal deoxynucleotidyl transferase-mediated nick end labeling and antifactor VIII antibody or antihuman NB demonstrated endothelial cell apoptosis in the vasculature, as well as increased tumor cell apoptosis (Pastorino *et al.*, 2003).

In another study by Lennernas *et al.* a tumor-free, biologically relevant model was used, as the influence of chemotherapy on tumor angiogenesis may be tumor and site dependent, and several variables of VEGF-mediated angiogenesis were studied. VEGF was injected intraperitoneally at picomolar concentrations, approximately physiologically relevant doses in rats. Various cytotoxic drugs, amongst which is DOX were subsequently administered in single intravenous injections at comparatively low, intermediate and high doses. Seven days later, the response in the mesentery was quantified in terms of microvessel spatial extension, density, pattern formation and microvessel segment length. Angiogenesis was significantly suppressed by doxorubicin at all three doses. The microvessel network pattern was significantly affected in some aspects. Intermediate and high (17 and 30 mg/kg, respectively) doses significantly decrease the vascularized area (by 46% and 66%, respectively) as compared to controls. The high dose also significantly reduced the microvascular length. Both the vascularized area and the microvascular length are measurements of microvessel proliferation. The number of branching points per unit tissue volume and the number of segments per unit tissue volume decreased significantly at the high dose. Both these reflect the pattern and proliferation of microvessels. Moreover, all doses of doxorubicin significantly lengthened the distance between two successive branching points, or the length of individual segments (Lennernas *et al.*, 2003).

2.3. Naphthoquinones

1,4-Naphthoquinones form a family of compounds characterized by a naphthalene ring containing two carbonyl moieties at positions 1 and 4 (Figure 3). Vitamin K is an important naphthoquinone; it is a cofactor in blood coagulation that acts by post-translational modification of a number of plasma proteins. Recent research has found that it is also involved in bone metabolism (Shearer, 2000). Moreover, several vitamin K-dependent proteins have been identified as ligands for receptor tyrosine kinases, which are an essential element in cell signaling pathways (Lamson and Plaza, 2003). Vitamin K has been shown to inhibit cancerous cell growth both *in vivo* and *in vitro*. This was explained by the redox-cycling that quinones undertake; nevertheless, evidence supporting a mechanism besides redox-cycling was provided when vitamin K3 was found to be still toxic after the addition of an iron chelator that would inhibit ROS formation by Fenton reactions (Cantoni *et al.*, 1991). It was shown that the binding of vitamin K3 to the catalytic domain of Cdc25 phosphatase results in the formation of inactive, hyperphosphorylated Cdk1, which subsequently induces cell cycle arrest, leading to cell death (Wu and Sun, 1999).

Yoshiji *et al.* have studied the effects of a combination treatment involving the clinically used vitamin K2 and angiotensin-converting enzyme inhibitor Perindopril on rat hepatocarcinogenesis. They showed that the combination markedly inhibited the development of preneoplastic lesions in association with suppression of neovascularization in the liver; it exerted a more potent inhibitory effect as compared with the single agent treatments. In addition, the *in vitro* study demonstrated that vitamin K and Perindopril inhibited the endothelial cell tubular formation. Vitamin K also suppressed the endothelial cell proliferation in a dose-dependent manner (Yoshiji *et al.,* 2005).

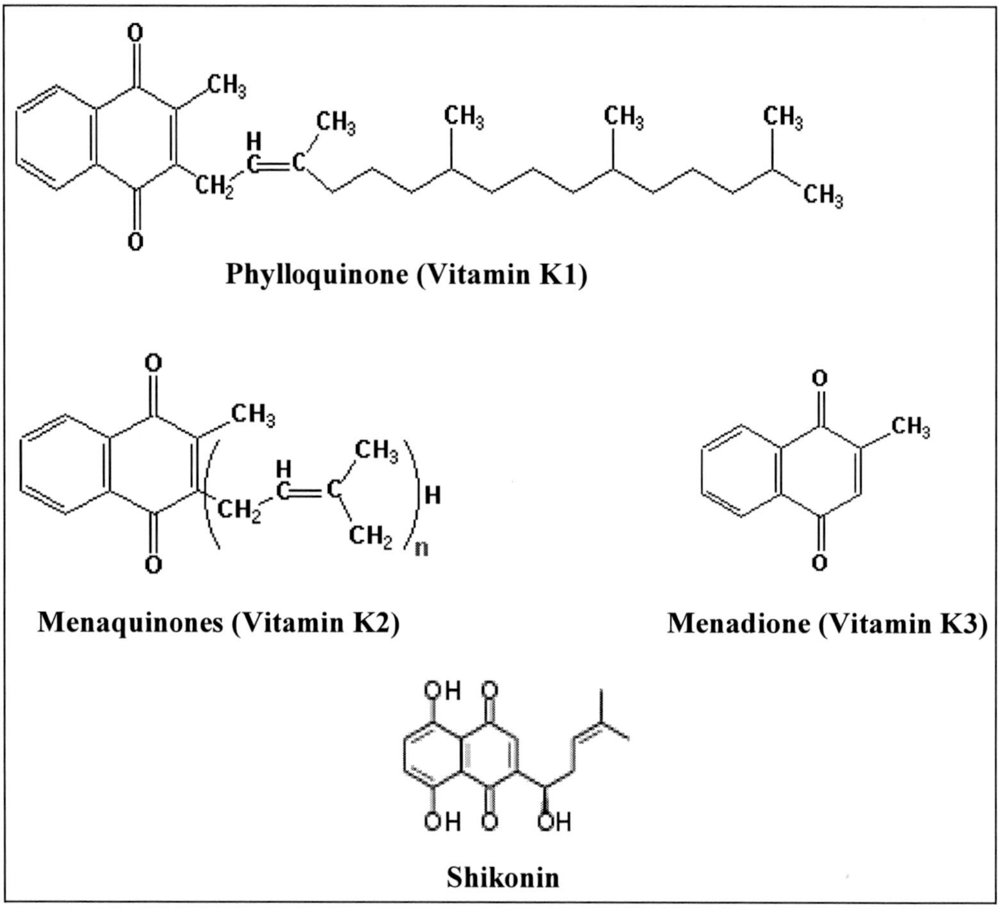

Figure 3. Structures of the Naphthoquinones, Phylloquinone, Menadione, Menaquinones (n= 6-10) and Shikonin

Shikonin is a naphthoquinone pigment isolated from *Lithospermum erythrorhizon* of the *Boraginaceae* family (Figure 3). Zicao, or the dried root of plants of this family, is a commonly used herbal medicine in China (Chen *et al.,* 2002). It contains a mixture of naphthoquinone pigments, and has been used for thousands of years for treatment of macular eruption, measles, burns, sore throat and cancer (Chen *et al.,* 2002). Shikonin showed complete tumor growth inhibition, and increased the life span of mice injected intraperitioneally with sarcoma S-180 cells (Chen *et al.,* 2002). Induction of apoptosis by shikonin in human colorectal carcinoma cell lines was through coordinative modulation of the Bcl-2 family, p27, and p53, release of cytochrome c, and sequential activation of caspases

(Hsu *et al.,* 2004). Another extract from Zicao of *Arnebia euchroma*, naphthoquinone pigment-L III, inhibited the proliferation of gastric and oesophagal cancer cell lines (Chen *et al.,* 2002). β-Hydroxyisovalcrylshikonin is another shikonin variant that inhibited growth of various cell lines including colon malignant melanoma, colon adenocarcinoma, and gastric and pancreatic cancer cell lines. The induction of apoptosis in these cell lines was due to the activation of caspase-3 pathway. A recent study of Shikonin conducted on hepatoma cell lines showed that a potential mode of action is that of ROS generation; pretreatment of shikonin-treated cells with the antioxidants glutathione and N-acetyl cysteine protected the cells from apoptosis (Chen *et al.,* 2003). In addition, Shikonin was found to be a promising chemopreventive agent for intestinal neoplasia; when added to the diet of rats (0.02%), it decreased the average number of tumors in rats treated with azoxymethane (Yoshimi *et al.,* 1992).

In a murine model, angiogenesis induced by TNF-α was inhibited when shikonin was injected along with TNF-α (Hisa *et al.,* 1998). When co-injected with B16 melanoma cells, shikonin strongly inhibited tumor growth and tumor-induced angiogenesis; histological analysis of the B16 cells suggested that shikonin's effects weren't due to cytotoxicity, but to prevention of neovascularization. This was also seen using the chick chorioallantoic membrane assay; shikonin doses that had no effect on the development of the embryo were able to inhibit angiogenesis. Also, shikonin prevented formation of capillary-like tubes by bovine aortic endothelial cells when seeded on Matrigel extracellular matrix (Hisa *et al.,* 1998).

3. Other Quinones with Antiangiogenic Activities

3.1. Thymoquinone

Thymoquinone (TQ), or 2-isopropyl-5-methyl-1,4-benzoquinone, is the main constituent of the volatile oil of *Nigella sativa* seeds (Figure 4). TQ showed significant antineoplastic activity against human pancreatic adenocarcinoma, uterine sarcoma, and leukemic cell lines, while it is minimally toxic to normal cells (Worthen *et al.*, 1998). The multidrug-resistant variants of human pancreatic adenocarcinoma, uterine sarcoma, and leukemic cell lines, which are over 10-fold more resistant to the standard antineoplastic agents doxorubicin and etoposide as compared to their respective parental controls, were equally sensitive to TQ (Worthen *et al.*, 1998). TQ has been shown to induce apoptosis by p53-dependent (Gali-Muhtasib *et al.*, 2004) and p53-independent (El-Mahdy *et al.*, 2005) pathways. In HCT-116 human colon cancer cells, the pro-apoptotic effects of TQ are dependent on p53 (Gali-Muhtasib *et al.*, 2004), while in myeloblastic leukemia HL-60 cells, TQ-induced apoptosis is p53-independent and occurs through the activation of caspase 8, 9, and 3 (El-Mahdy *et al.*, 2005). TQ treatment of HCT-116 cells resulted in a marked increase in p53 and p21 protein levels, and in a significant inhibition of the anti-apoptotic Bcl-2 protein (Gali-Muhtasib *et al.*, 2004). Co-incubation with the specific p53-inhibitor, pifithrin-alpha, restored the mRNA and protein levels of Bcl-2, p53, and p21 to control levels and suppressed TQ-induced apoptosis.

Recent studies in our laboratory have shown that TQ decreases the protein levels of HIF-1α in DLD-1 and Caco-2 human colon carcinoma cells exposed to the hypoxia mimic Cobalt Chloride (Gali-Muhtasib *et al.*, unpublished findings). Others have also shown that TQ prevents the secretion of VEGF, DNA-binding activity of HIF-1α, and cancer cell migration in a variety of cell lines, including Caco-2 and MCF-7 human breast cancer cells (reviewed in Gali-Muhtasib *et al.*, 2006).

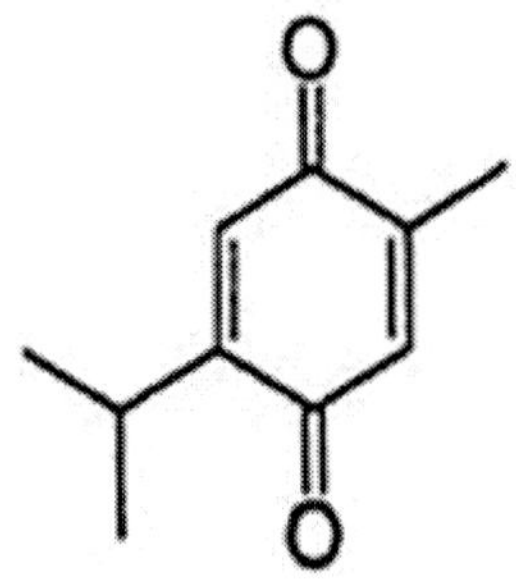

Figure 4. Thymoquinone

3.2. Cochlioquinone A1

While screening for new antiangiogenic agents derived from natural products, Jung *et al.* isolated cochlioquinone A1 (CoA1) from the culture extract of *Bipolaris zeicola* fungal strain. CoA1 is identified as a derivative of cochlioquinone A (CoA) (Figure 5). CoA1 contains an acetone moiety added to one of the quinone carbonyl groups found in CoA and its analogues. It was shown that CoA1 potently and selectively inhibited the growth of bovine aortic endothelial cells rather than cancer cells. It also inhibited the Fibroblast Growth Factor (FGF)-induced invasion of bovine aortic endothelial cells and tube formation on Matrigel in a dose-dependent manner, without affecting the viability of the cells (Jung *et al.*, 2003).

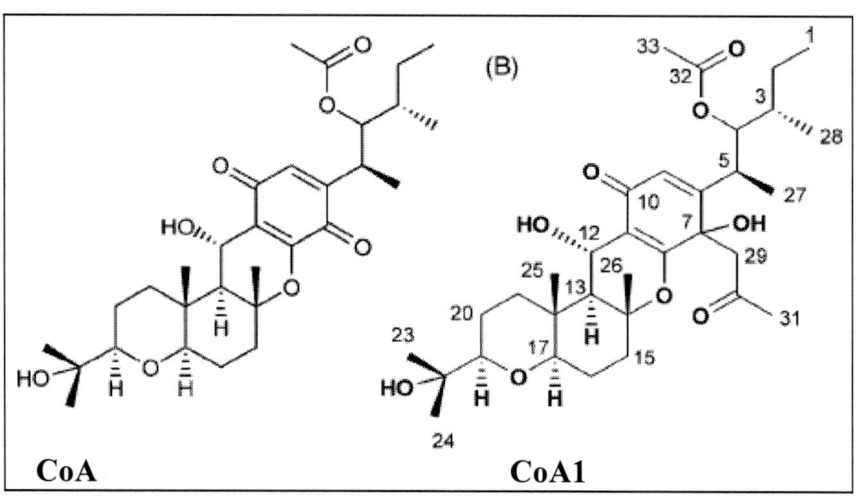

Figure 5. Cochlioquinones

3.3. Cannabidiol Hydroxyquinone: HU-331

In 2004, Kogan *et al.* reported the synthesis of the new anticancer quinine, cannabidiol hydroxyquinone (HU-331), from cannabidiol, one of the most abundant cannabinoids of *Cannabis sativa* (Figure 6). HU-331 inhibited the growth of various cell lines including Burkitt's lymphoma (Raji), T-cell lymphoma (Jurkat), glioblastoma (SNB-19), breast cancer (MCF-7), prostate cancer (DU-145), lung cancer (NCI-H-226), and colon cancer (HT-29). The compound was also highly effective against tumor xenografts in nude mice (Kogan *et al.*, 2004). A possible role attributed to cannabinoids with regards to inhibition of angiogenesis is their ability to modulate the response of cells to growth factors such as VEGF and FGF. In fact, numerous cannabinoids inhibit endothelial cell survival and migration (Blazquez *et al.*, 2003). In particular, HU-331 was able to induce apoptosis of bovine aortic endothelial cells, and was strongly antiangiogenic in an *ex vivo* rat aortic ring assay at low doses: the area of endothelial cells that proliferate from aorta, the number of new vessels formed, the average and maximal vessel lengths were all decreased (Kogan *et al.*, 2006). In the *in vivo* model of whereby HT-29 cells are inoculated in mice, HU-331 significantly decreased the number of blood vessels per area unit. In addition, transcription of genes related to various basic functions of vascular endothelial cells were modified by HU-331: the mRNA levels of matrix metalloproteinase-1 (MMP-1), cyclooxygenase-2 (COX-2) and osteoprotegerin (OPG) were increased, while the mRNA levels of monocyte chemotactic protein-1 (MCP-1), von Willebrand factor (VWF) and cytosolic phospholipase A2 (PLA2) were decreased.

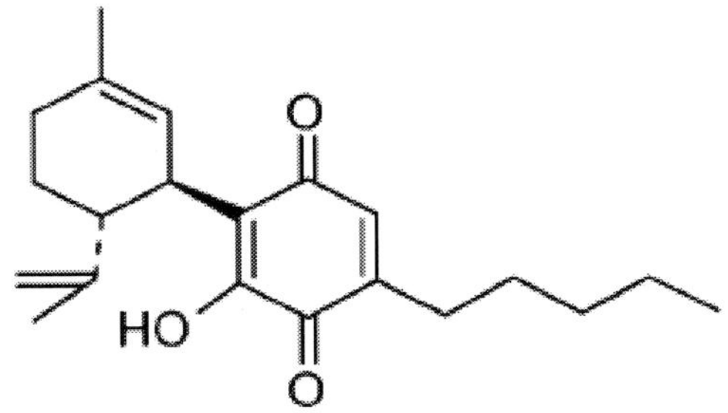

Figure 6. Cannabidiol Hydroxyquinone (HU-331)

3.4. Linomide

Linomide is a quinone-3-carboxamide that has shown antitumor activity against a series of androgen-dependent and androgen–independent rat prostate cancers, as well as human prostate carcinoma (Figure 7). Interestingly, the compound has the ability to inhibit the growth of prostatic cancer *in vivo* but not *in vitro* (Ichikawa *et al.*, 1992). The reason for this discrepancy is that linomide inhibits tumor growth not directly but indirectly *in vivo* via its ability to inhibit the angiogenic response induced within the growing prostatic cancer. The carboxamide's antiangiogenic activity is believed to be due to its ability to inhibit endothelial cell migration and invasion via inhibiting TNF-α synthesis and secretion by tumor-associated macrophages. In fact, treatment of rats with linomide decreases significantly the number of tumor-associated macrophages within both locally invasive and highly metastatic primary prostatic cancers. Monocytes/macrophages isolated from linomide treated rats had a decreased ability to secrete TNF-α when challenged *in vitro* with the bacterial endotoxin lipopolysaccharide (Vukanovic and Isaacs, 1995; Gourley and Williamson, 2000).

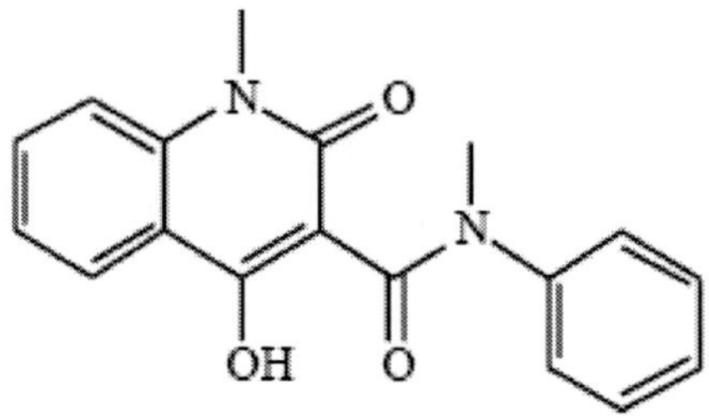

Figure 7. Linomide

4. Conclusion

As angiogenesis is a crucial step in tumor metastasis, antiangiogenic therapy is a central key for the successful combination treatment of any kind of tumor. As established antineoplastic agents, we reviewed in this paper the antiangiogenic potential of anticancer quinones, and found differences in the advances made for different classes. For instance, the molecular pathway by which Benzoquinone ansamycins inhibit angiogenesis are well-established, and some ansamycins have entered clinical trials. More research is crucial for the precise determination of the mode of action of other quinones. Such research is imperative for any of these compounds to enter clinical trials. Indeed, by identifying the chemical elements responsible for the antiangiogenic effect, more successful derivatives of these quinones would hopefully be developed.

References

Asche C. (2005). Antitumor quinones. *Mini Reviews in Medicinal Chemistry*, *5(5)*, 449-67.

Blazquez, C., Casanova, M.L., Planas, A., Del, Pulgar. T.G., Villanueva, C., Fernandez-Acenero, M.J., Aragones, J., Huffman, J.W., Jorcano, J.L. and Guzman, M. (2003). Inhibition of tumor angiogenesis by cannabinoids. *Federation of American Societies for Experimental Biology Journal*, *17(3)*, 529-531.

Burridge, K., Turner, C.E. and Romer, L.H. (1992). Tyrosine phosphorylation of paxillin and 125FAK accompanies cell adhesion to extracellular matrix: A role in cytoskeletal assembly. *Journal of Cell Biology*, *119*, 893–903.

Cantoni, O., Fiorani, M., Cattabeni, F. and Bellomo, G. (1991). DNA breakage caused by hydrogen peroxide produced during the metabolism of 2-methyl- 1,4-naphthoquinone (menadione) does not contribute to the cytotoxic action of the quinone. *Biochemical Pharmacology*, *42*, S220-S222.

Chen, C.H., Chern, C.L., Lin, C.C., Lu, F.J., Shih, M.K., Hsieh, P.Y. and Liu, T.Z. (2003). Involvement of reactive oxygen species, but not mitochondrial permeability transition in the apoptotic induction of human SK-Hep-1 hepatoma cells by shikonin. *Planta Medica, 69(12)*, 1119-1124.

Chen, X., Yang, L., Oppenheim, J.J. and Howard, M.Z. (2002). Cellular pharmacology studies of shikonin derivatives. *Phytotherapy Research, 16(3)*, 199-209.

El-Mahdy, M. A., Zhu, Q., Wang, Q. E., Wani, G. and Wani AA. (2005). Thymoquinone induces apoptosis through activation of caspase-8 and mitochondrial events in p53-null myeloblastic leukemia HL-60 cells. *International Journal of Cancer, 117*, 409-417.

Fernandez, A.M., Van Derpoorten, K., Dasnois, L., Lebtahi, K., Dubois, V., Lobl, T.J., Gangwar, S., Oliyai, C., Lewis, E.R., Shochat, D. and Trouet, A. (2001). N-Succinyl-(beta-alanyl-L-leucyl-L-alanyl-L-leucyl)doxorubicin: an extracellularly tumor-activated prodrug devoid of intravenous acute toxicity. *Journal of Medicinal Chemistry, 44(22)*, 3750-3753.

Folkman, J. and Shing, Y. (1992). Angiogenesis. *The Journal of Biological Chemistry, 267(16)*, 10931-10934.

Gali-Muhtasib, H., Diab-Assaf, M., Boltze, C., Al-Hmaira, J., Hartig, R., Roessner, A. and Schneider-Stock, R. (2004). Thymoquinone extracted from black seed triggers apoptotic cell death in human colorectal cancer cells via a p53-dependent mechanism. *International Journal of Oncology, 25*, 857-866.

Gali-Muhtasib, H., Roessner, A. and Schneider-Stock, R. (2006). Thymoquinone: A promising anti-cancer drug from natural sources. *The International Journal of Biochemistry and Cell Biology, 38(8)*, 1249-1253.

Goetz, M.P., Toft, D.O., Ames, M.M. and Erlichman, C. (2003). The Hsp90 chaperone complex as a novel target for cancer therapy. *Annals of Oncology, 14(8)*, 1169-1176.

Gourley, M. and Williamson, J. (2000). Angiogenesis: New Targets for the Development of Anticancer Chemotherapies. *Current Pharmaceutical Design, 6*, 417-439.

Hisa, T., Kimura, Y., Takada, K., Suzuki, F. and Takigawa, M. (1998). Shikonin, an ingredient of Lithospermum erythrorhizon, inhibits angiogenesis in vivo and in vitro. *Anticancer Research, 18(2A)*, 783-790.

Hsu, P.C., Huang, Y.T., Tsai, M.L., Wang, Y.J., Lin, J.K. and Pan, M.H. (2004). Induction of apoptosis by shikonin through coordinative modulation of the Bcl-2 family, p27, and p53, release of cytochrome c, and sequential activation of caspases in human colorectal carcinoma cells. *Journal of Agriculture and Food Chemistry, 52(20)*, 6330-6337.

Ichikawa, T., Lamb, J.C., Christensson, P.I., Hartley-Asp, B. and Isaacs, J.T. (1992). The antitumor effects of the quinoline-3-carboxamide linomide on Dunning R-3327 rat prostatic cancers. *Cancer Research, 52(11)*, 3022-3028.

Jung, H.J., Lee, H.B., Lim, C., Kimb, C. and Kwona, H.J. (2003). Cochlioquinone A1, a New Anti-Angiogenic Agent from Bipolaris zeicola. *Bioorganic and Medicinal Chemistry, 11*, 4743–4747.

Kleihues, P. and Cavenee, W.K. (2000). World Health Organization Classification of Tumors: Pathology and genetics of tumors of the nervous system. Lyon: IARC Press.

Kogan, N., Blazquez, C., Alvarez, L., Gallily, R., Schlesinger, M., Guzman, M. and Mechoulam, R. (2006). A cannabinoid quinone inhibits angiogenesis by targeting vascular endothelial cells. *Molecular Pharmacology, 70(1),* 51-59.

Kogan, N., Rabinowitz, R., Levi, P., Gibson, D., Sandor, P., Schlesinger, M. and Mechoulam, R. (2004). Synthesis and Antitumor Activity of Quinonoid Derivatives of Cannabinoids. *Journal of Medicinal Chemistry*, *47*, 3800-3806.

Lamson, D and Plaza, S. (2003). The Anticancer Effects of Vitamin K. *Alternative Medicine Review, 8(3)*, 303-318.

Lennernas, B., Albertsson, P., Lennernas, H. and Norrby, K. (2003). Chemotherapy and antiangiogenesis--drug-specific, dose-related effects. *Acta Oncologica*, *42(4)*, 294-303.

Mabjeesh, N., Post, D., Willard, M., Kaur, B., Van Meir, E., Simons, J. and Zhong, H. (2002). Geldanamycin Induces Degradation of Hypoxia-inducible Factor 1α Protein via the Proteosome Pathway in Prostate Cancer Cells. *Cancer Research, 62(9)*, 2478–2482.

Minotti, G., Menna, P., Salvatorelli, E., Cairo, G. and Gianni, L. (2004). Anthracyclines: molecular advances and pharmacologic developments in antitumor activity and cardiotoxicity. *Pharmacological Reviews*, *56(2)*, 185-229.

Murakami, Y., Mizuno, S., Hori, M. and Uehara, Y. (1988). Reversal of transformed phenotypes by herbimycin A in src oncogene expressed rat fibroblasts. *Cancer Research*, *48(6)*, 1587-1590.

Neckers, L. and Neckers, K. (2002). Heat-shock protein 90 inhibitors as novel cancer chemotherapeutic agents. *Expert Opinion on Emerging Drugs*, *7(2)*, 277-288.

O'Brien, P.J. (1991). Molecular Mechanisms of Quinone Cytotoxicity. *Chemico-Biological Interactions*, *80*, 1-41.

Pastorino, F., Brignole, C., Marimpietri, D., Cilli, M., Gambini, C., Ribatti, D., Longhi, R., Allen, T., Corti A. and Ponzoni, M. (2003). Vascular Damage and Anti-angiogenic Effects of Tumor Vessel-Targeted Liposomal Chemotherapy. *Cancer Research*, *63*, 7400-7409.

Rousseau, S., Houle, F., Kotanides, H., Witte, L., Waltenberger, J., Landry, J. and Huot, J. (2000). Vascular endothelial growth factor (VEGF)-driven actin-based motility is mediated by VEGFR2 and requires concerted activation of stress-activated protein kinase 2 (SAPK2/p38) and geldanamycin-sensitive phosphorylation of focal adhesion kinase. *The Journal of Biological Chemistry, 275*, 10661–10672.

Sanderson, S., Valenti, M., Gowan, S., Patterson, L., Ahmad, Z., Workman, P. and Eccles, S.A. (2006). Benzoquinone ansamycin heat shock protein 90 inhibitors modulate multiple functions required for tumor angiogenesis. *Molecular Cancer Therapeutics*, *5(3)*, 522-532.

Semenza, G. L. (2002). HIF-1 and tumor progression: pathophysiology and therapeutics. *Trends in Molecular Medicine*, *8*, S62-S67.

Shearer, M.J. (2000). Role of vitamin K and Gla proteins in the pathophysiology of osteoporosis and vascular calcification. *Current Opinion in Clinical Nutrition and Metabolic Care*, *3*, 433-438.

Uehara, Y., Hori, M., Takeuchi, T. and Umezawa H. (1986). Phenotypic change from transformed to normal induced by benzoquinonoid ansamycins accompanies inactivation of p60src in rat kidney cells infected with Rous sarcoma virus. *Molecular and Cellular Biology 6(6)*, 2198-2206.

Vukanovic, J. and Isaacs, J.T. (1995). Human prostatic cancer cells are sensitive to programmed (apoptotic) death induced by the antiangiogenic agent linomide. *Cancer Research, 55(16)*, 3517-3520.

Weiss, R.B. (1992). The anthracyclines: will we ever find a better doxorubicin? *Seminars in Oncology*, 19(6), 670-686.

Whitesell, L., Mimnaugh, E.G., De Costa, B., Myers, C.E. and Neckers L.M. (1994). Inhibition of heat shock protein HSP90-pp60v-src heteroprotein complex formation by benzoquinone ansamycins: essential role for stress proteins in oncogenic transformation. *Proceedings of the National Academy of Sciences of the United States of America, 91(18)*, 8324-8328.

Whitesell, L., Shifrin, S.D., Schwab, G. and Neckers, L.M. (1992). Benzoquinonoid ansamycins possess selective tumoricidal activity unrelated to src kinase inhibition. *Cancer Research, 52(7)*, 1721-1728.

Worthen, D. R., Ghosheh, O. A. and Crooks, P. A. (1998). The in vitro anti-tumor activity of some crude and purified components of blackseed, Nigella sativa L. *Anticancer Research, 18*, 1527-1532.

Wu, F.Y. and Sun, T.P. (1999). Vitamin K3 induces cell cycle arrest and cell death by inhibiting CDC25 phosphatase. *European Journal of Cancer, 35*, 1388-1393.

Yoshiji, H., Kuriyama, S., Noguchi, R., Yoshii, J., Ikenaka, Y., Yanase, K., Namisaki, T., Kitade, M., Yamazaki, M., Masaki, T. and Fukui, H. (2005). Combination of vitamin K2 and the angiotensin-converting enzyme inhibitor, perindopril, attenuates the liver enzyme-altered preneoplastic lesions in rats via angiogenesis suppression. *Journal of Hepatology, 42(5)*, 687-693.

Yoshimi, N., Wang, A., Morishita, Y., Tanaka, T., Sugie, S., Kawai, K., Yamahara, J. and Mori H. (1992). Modifying effects of fungal and herb metabolites on azoxymethane-induced intestinal carcinogenesis in rats. *Japanese Journal of Cancer Research, 83(12)*, 1273-1278.

Zagzag, D., Nomura, M., Friedlander, D.R., Blanco, C.Y., Gagner, J.P., Nomura, N. and Newcomb, E.W. (2003). Geldanamycin inhibits migration of glioma cells in vitro: a potential role for hypoxia-inducible factor (HIF-1alpha) in glioma cell invasion. *Journal of Cellular Physiology, 196(2)*, 394-402.

In: New Research on Angiogenesis Inhibitors
Editor: Elmer T. Skinard, pp. 115-133

ISBN: 1-60021-473-8

Chapter VI

BEVACIZUMAB (AVASTIN) IN NEOVASCULAR OCULAR DISEASE

Nathan J. Rudometkin[1,2], Gian Paolo Giuliari[1,2] and Antonio P Ciardella[2,*]

[1]University of Colorado Department of Ophthalmology, Rocky Mountain Lions Eye Institute, Aurora CO; USA

[2]Denver Health Hospital Authority, Department of Ophthalmology, Denver CO.USA

ABSTRACT

Angiogenesis is currently one of the most important topics in medicine. The regulation and control of blood vessel growth is important for both normal and diseased tissues. Vascular endothelial growth factor (VEGF) with its various subtypes is a vital signal in the pathway of vascular proliferation. Expression of VEGF has been noted in multiple types of cancer and in ocular diseases with vascular proliferation. Bevacizumab (Avastin) is a potent monoclonal antibody against all isoforms of VEGF that has been successfully used to treat colorectal, breast and lung cancer. The success of a related medicine, ranibizumab for treating neovascularization age related macular degeneration (AMD), prompted investigators to use bevacizumab "off-label" for neovascular AMD. Promising results have been found in treating exudative AMD both intravenously and more commonly through direct intravitreal injection with impressive improvements in patient's vision, ocular coherence tomography (OCT) and angiographic findings. Bevacizumab in both animal and human subjects appears to be safe and well tolerated in short-term evaluation. This antibody has been shown to penetrate extensively throughout the retina in albino rabbits. Other small series report successful treatment of proliferative diabetic retinopathy and retinal venous occlusive disease.

* Correspondence concerning this article should be addressed to Dr. Antonio Ciardella MD, Chief Department of Ophthalmology, Denver Health Hospital Authority, 777 Bannock Street, MC 0156 Denver, CO 80204. Telephone: 303 436-3448; Fax: 303 436-3181; Email: aciardella@yahoo.com.

Over seven thousand intravitreal injections of bevacizumab have been administered in twelve countries as reported in the Avastin Safety Survey. To this point bevacizumab appears to be a safe and efficacious treatment of neovascular AMD and possibly other retinal vascular diseases. Despite the reported short-term success large controlled trials are necessary to clearly define the optimal treatment protocols and the precise frequency of adverse events. Consequently, care must be taken when treating patients with bevacizumab with detailed discussions about the possible adverse events of this "off-label" treatment. Currently, intravitreal bevacizumab appears to be a promising and cost-effective treatment for a variety of important ocular diseases.

INTRODUCTION

Every tissue in the body requires the presence of blood vessels for development and repair. New blood vessel development begins with differentiation of angioblasts, which are the precursors to vascular endothelial cells [1]. Angiogenesis consists of reorganization and pruning of these endothelial cells that is essential for the formation and maintenance of normal vasculature [2]. Angiogenesis also plays an important role in the pathogenesis of different types of systemic diseases including many types of cancers, manifestations of rheumatoid arthritis and psoriasis, and in ocular diseases such as proliferative retinopathies and age related macular degeneration [3].

History

Theories have existed for many years suggesting a factor that stimulates and regulates angiogenesis and was proposed early in conjunction with eye disease. Michaelson hypothesized in 1948 that a factor X [4] synthesized and secreted from ischemic retina was causal for pathologic neovascularization in the retina, iris and optic nerve. Later, Judah Folkman presented the idea of inhibiting angiogenesis as a means of treating cancer in 1971 and spearheaded the efforts to discover tumor angiogenesis factors [5]. The identification of a key mediators of angiogenesis in 1983 with Senger, Dvorak, and colleagues who discovered a protein secreted from a guinea-pig tumor cell line that was a potent inducer of vascular leakage and named it vascular permeability factor (VPF) [6]. In 1989, Napoleone Ferrara and colleagues identified a molecule in media conditioned from bovine pituitary follicular cells that promoted the proliferation of endothelial cells and termed it vascular endothelial growth factor (VEGF) [7]. Ultimately, the cloning of VPF by Daniel Connelly and coworkers and VEGF by Ferrara's group demonstrated that the two factors were the same protein [8,9]. Forty years after Michaelson's initial proposal of "factor X" produced by the retina the molecule named VEGF was finally isolated.

VASCULAR ENDOTHELIAL GROWTH FACTOR (VEGF)

There are different types of VEGF including: VEGF A, VEGF B [10], VEGF C [11], VEGF D [12], VEGF E and placental growth factor (PIGF) all derived from the same gene family [13]. Our discussion will center on VEGF A since it appears to be involved in vascular proliferation whereas VEGF C and D relate to lymphatic proliferation [14]. VEGF B and PIGF are also involved in angiogenesis, but less is know about their exact role.

The family of VEGF-A proteins is derived from the same gene composed of eight exons, separated by seven introns [15,16]. Alternative exon splicing results in the generation of the four principal isoforms (VEGF 121, 165, 189, 206) named based on the number of amino acids present following signal-sequence cleavage [9]. Most isoforms bind heparin to some degree, except $VEGF_{121}$. Heparin binding relates to the ability of VEGF to bind to cell surfaces and basement membranes with the larger isoforms having increased heparin binding. Consequently, $VEGF_{121}$ can diffuse freely through cell membranes with the predominant isoform $VEGF_{165}$ having an intermediate level of heparin binding.

There are two receptors that VEGF has high-affinity, VEGF Receptor 1 (VEGFR-1) [17] and VEGF Receptor 2 (VEGFR-2) [18] with these receptors being located primarily on vascular endothelial cells [19]. The VEGFR-2 seems to be the major mediator of the angiogenic and permeability enhancing effects of VEGF [20].

The primary sources of VEGF in the eye are neural cells, including ganglion, Muller, and retinal pigment epithelium cells [21]. These cell types increase VEGF production when subjected to hypoxia [22,23]. In pathologic retinal conditions elevated VEGF has been found in patients with AMD from surgically excised membranes [24,25] and in the aqueous and vitreous in patients with diabetic retinopathy and retinal vein occlusion [26].

VEGF and Tumor Cells

Inhibiting VEGF was initially studied in relation to the treatment of cancer. Since most tumors need to develop an additional blood supply for continued growth anti-VEGF medications were proposed as a new approach to cancer treatment. The ability of the tumor cells to promote vascular proliferation gives them a growth advantage relative to normal cells otherwise they may remain dormant for years [27]. Multiple tumor cell lines secrete VEGF in vitro, suggesting the possibility that this molecule may be a mediator of tumor angiogenesis. Furthermore, VEGF mRNA has been isolated in most human tumors studied, including: carcinoma of the lung [28,29], breast [30,31], gastrointestinal tract [32,33], bladder and kidney [34], ovary [35,36], endometrium [37], several intracranial tumors such as glioblastoma multiforme, sporadic and von Hippel-Lindau syndrome associated capillary hemangioblastoma [38,39], and endocrine tumors [40]. Blocking VEGF in transgenic mice with anti-VEGF monoclonal antibodies had an inhibitory effect on several tumor cell lines giving further support to anti-VEGF treatments in patients with cancer [41].

BEVACIZUMAB IN SYSTEMIC CANCER TREATMENT

Bevacizumab (Avastin) is the first extensively studied class of innovative drugs investigated in clinical studies used as adjunctive therapy to standard cancer treatment. It is a humanized antibody (93% human, 7% murin) against all isoforms of VEGF-A. The effect of this therapy appears to interfere with the growth of new blood vessels, which provide nutrients to the tumor and decrease the tumor interstitial pressure that may increase delivery of chemotherapy despite the decrease in vasculature to the tumor [42].

Bevacizumab is currently approved by the Food and Drug Administration (FDA) as first line therapy (February 26, 2004) for metastatic colorectal cancer when administered with fluorouracil and leucovorin [43]. A 4.7-month increase in median survival was shown when bevacizumab was added to the previously two mention chemotherapy agents than when the two agents were given alone. A phase III (E2100) clinical trial in 688 patients with chemonaive metastatic breast cancer showed preliminarily an overall 3-month increased survival and 5 month increased progression free survival in all subgroups (in a subsequent update) with bevacizumab (10 mg/kg) added to paclitaxel verses paclitaxel alone [44]. Results reported for the Phase III E4599 trial with 878 patients with non-squamous non-small cell lung cancer showed an improved median survival from 10.2 to 12.5 months for patients treated with the addition of bevacizumab (15 mg/kg) to paclitaxel and carboplatin verses these agents alone [45]. Application for FDA approval for bevacizumab to be used in patients with breast cancer and non-squamous non-small cell carcinoma is expected with the decision likely at the end of 2006. In summary, Avastin has proven to be a major breakthrough in cancer treatment. It is the first anti-cancer medication that significantly improves patient survival.

Description:

- Monoclonal antibody IgG1
- Weight ~149 kilodaltons
- pH 6.2
- Clear to slightly opalescent, colorless to pale brown, sterile solution
- Produced in a Chinese Hamster Ovary mammalian cell expression system in a nutrient medium containing the antibiotic gentamicin
- Supplied in 100 mg and 400 mg preservative-free, single-use vials to deliver 4 ml or 16 ml
- 100 mg formulation: 240 mg α,α-trehalose dihydrate, 23.2 mg sodium phosphate (monobasic, monohydrate), 4.8 mg sodium phosphate (dibasic, anhydrous), 1.6 mg polysorbate 20, and Water for Injection, USP

Pharmacokinetics:

- Half-life 20 days (range 11-50 days)*
- Estimated time to steady state 100 days*
- Accumulation ratio 2.8 (after 10 mg/kg every 2 weeks)
- Clearance: males = 0.262 L/day, females = 0.207 L/day
- V_c**: males = 3.25 L, females = 2.66 L
- Higher than median tumor burden yielded higher clearance

*based on total antibody concentration both free and ligand bound

**V_c = volume of distribution

Bevacizumab and Adverse Events

The adverse side effects are an important concern as these effects may be seen when used in patients with ocular disease. Safety was first evaluated in nonhuman primates (*Macaca fascicularis*). After administration for 4 to 13 weeks at doses up to 50 mg per kg, young adult cynomolgus monkeys had physeal dysplasia with dose related increases in hypertrophied chondrocytes and inhibition of vascular invasion of the growth plate. Also suppression of angiogenesis in the female reproductive tract was observed, resulting in decreased ovarian and uterine weights, and an absence of the corpora lutea. Both the growth-plate and ovarian changes were reversible with cessation of treatment. No other treatment related effects were observed [46].

Initially, multiple types of serious adverse events were being reported in patients using systemic bevacizumab based on data from the early clinical trials. At this time the adverse events are becoming better defined with results being reported from the larger Phase III trials (see Table 1) [43-45,47-50]. Grade 3/4 hypertension appears to be a consistent feature occurring in approximately 8-18% of patients, but is almost always well managed medically. Minor mucosal bleeding (gum and nose bleeding) and mild Grade 1/2 proteinuria appear to be relatively common with Grade 3/4 proteinuria occurring rarely at approximately 2% in one study. Neutropenia was seen in 24% of treated patients verses 17% of controls in one study and seven patients in the treatment and two controls developed congestive heart failure (CHF) in another report and although not statistically significant, development or worsening of CHF may be a concern. Life treating hemorrhagic events (Grade 3/4) occurred rarely (approximately 2-5%) in some but not all of the Phase III trials with severe pulmonary hemorrhage seen in patients with non-small cell lung cancer being an important finding. Thrombosis conversely, including stroke, myocardial infarction (MI), and peripheral arterial thrombosis initially a concern in the smaller phase II trials has not been significantly increased in the larger Phase III trials. Although most of both the Phase II and III trials excluded patients with any history of prior stroke or uncontrolled cardiovascular disease within the last 6 months (MI, unstable angina, uncontrolled hypertension, > Grade 2 CHF). So, concern may arise in treating patients in these particular groups that were excluded.

Gastrointestinal perforation was an important but rare event (1.5%) in patients with colorectal cancer. In this study two of the six patients had prior colon surgery suggesting a concern with bevacizumab and impaired wound healing. Since, angiogenesis may play a role in wound reanastomosis and blocking this process may increase postoperative complications. A recent review examined data from two large studies to determine the risk of bevacizumab in association with recent surgery [51]. No significant increase in wound healing complications were seen in patients receiving bevacizumab 28-60 days after surgery (1.3% bevacizumab + chemotherapy verses 0.5% chemotherapy alone), but in those who were receiving bevacizumab at the time of surgery a nonsignificant increase in wound healing complications 13% verses 3% of controls was observed. Close monitoring and consideration was suggested for patients receiving bevacizumab at the time of surgery.

Also, certain patients were excluded from the studies discussed above and increase in adverse event rates may occur if treating patients in these groups. The common exclusion criteria used in the studies mentioned previously include: major surgery four weeks prior to initiation of treatment, clinically significant cardiovascular disease (defined above), pregnancy or lactation, regular aspirin use (>325 mg per day) or other nonsteroidal anti-inflammatory agents, preexisting bleeding diatheses, coagulapathy or need for full dose anticoagulation. These factors should be considered when treating patients' ocular disease with bevacizumab.

Table 1. Systemic Bevacizumab and Adverse Events in Cancer Treatment.

Study	Phase	N=	Dosing (mg/ Kg)	Dosing Interval/ Total Treatment Duration (in weeks)	Minor bleeding Grade 1/2 (tx/control)	Proteinuria Grade 1/2 (tx/control)	Hypertension Grade 3/4 (tx/control)	Hemorrhagic Events Grade 3/4 (tx/control)	Thrombotic Events Grade 3/4 (tx/control)	Miscel-laneous
Advanced or Metastatic Lung Cancer (E4599) [45]	III	878	15	3/24	Not currently reported	Not currently reported	8%/<1%*	4.7%/1.1%⇀ (2.3% pulmonary hemorrhage)	3%/3.8%	Neutropenia Grade 4 24%/17%* (tx/control)
Metastatic Breast Cancer (E2100) [44]	III	680	10	2	Not currently reported	2%/0%* (Grade 3/4)	16%/2%*	2%/0%	2%/4%	Neuropathy (tx/control) 20%/14%⇀
Metastatic Breast Cancer [47]	III	444	15	3/105	28.4%/10.7%*	21.4%/7.4%*	17.9%/0.5%*	0.4%/0.5%	5.6%/3.7 %	CHF development in nine (7 tx/2 control)
Colorectal Cancer [43]	III	813	5	2/96	Not reported	25.7%/20.9%	11.0%/2.3%*	3.1%/2.5%	19.4%/16.2%	Six GI perforation (1.5%) in tx group, one case fatal
Colorectal Cancer [48]	II	204	5	2/96	Not reported	37%/19%*	16%/3%*	5%/3%	18%/18%	Arterial thrombosis (tx/control) 10%/5%⇀; Two GI perforation in tx group
Colorectal Cancer [49]	II	104	5 or 10	2/48	49.3%/11%⇀	25%/11.4%⇀ (grade not specified)	16.4%/0%⇀	4.5%/0%⇀	10.4%/2.9%⇀	
Advanced or Metastatic Lung Cancer [50]	II	99	7.5 or 15	3/18	37.9%/6.3%⇀	31.8%/6.3%⇀	3.0%/3.1%⇀	9.0%/0%⇀	10.6%/9.4%⇀	Six life threatening pulmonary hemorrhages

References in brackets [x], where x = reference number; * = represent significant differences; ⇀ = not specifically discussed if differences significant; tx = treatment; GI = gastrointestinal; CHF = congestive heart failure

Bevacizumab and Ocular disease

Ocular Safety and Penetration

When considering the safety of treating patients with intravitreal bevacizumab (1.25 mg in 0.05 ml) in relation to the previously mentioned adverse events found with systemic treatment several factors should be mentioned. The patients being treated with advanced or metastatic cancer are receiving a systemic dose approximately 350-1000 times greater (5-15 mg/Kg) than the intravitreal dose every 2-3 weeks for several months and for most medicines toxicities are dose related. It would be suspected that minimal amounts of bevacizumab would diffuse into the systemic circulation after an intravitreal injection. But these concentrations would be extremely low and a report citing personal communication with the manufacture stated that current available assays could not detect the systemic levels expected given an intravitreal injection [52]. But certainly these serious adverse events reported for systemic treatment should be a concern and carefully weighed in discussions with patients who may undergo an intravitreal injection until long-term controlled data is available on intravitreal bevacizumab.

Several studies have evaluated the short-term potential toxicity of bevacizumab in both animal and human studies. In one report twelve New Zealand albino rabbits were injected with four different concentrations (0.5, 1.0, 2.5 per 0.1 ml and 5.0 mg per 0.2 ml) with the fellow eye injected with 0.1 ml of balanced saline solution [53]. The eyes were observed for two weeks with slit-lamp and funduscopic examination, electroretinograms (ERG) prior and two weeks after the injections. Eyes underwent histological examination with hematoxylin and eosin staining for light microscopy evaluation. The only significant findings observed was a 11% decline in ERG amplitudes in one of the three eyes injected with the 5.0 mg dose. Declines were only considered significant if they were greater than 30% and one of the control eyes showed a 13% decrease in amplitude. Light microscopy reveled no signs of toxicity apart from some inflammatory cells in the vitreous in one eye injected with the 5.0 mg dose. Another study evaluated eight Dutch-belted rabbits, half injected with 1.25 mg/0.05 ml or 2.5 mg/1.0 ml of bevacizumab with the fellow eye serving as the control [54]. After one month no signs of toxicity were seen using light microscopy.

Another report evaluated intravitreal bevacizumab injected (2.5 mg/0.1 ml) into ten albino rabbits with saline control of the fellow eye [55]. ERG responses were evaluated at three hours, three days, and one, two and four weeks post injection without significant changes in the amplitudes. Visual evoked potential (VEP) also showed no abnormalities when recorded at four weeks. Evaluation with immunohistochemistry showed retinal penetration of bevacizumab throughout all layers of the retina at 24 hours. Muller cell labeling was increased after one week, but no other significant labeling was found in the inner and outer segment layers and at four weeks no labeling in the retina was observed until the gain was significantly increased at which time faint staining of the internal limiting membrane was present. Early reports suggesting that bevacizumab was too large to penetrate the retina based on studies with an analogous molecule herceptin (an early anti-VEGF antibody) appear to be incorrect [56] and no significant retinal toxicity was seen in these animal studies.

Nine human subjects had both multifocal electroretinograhy (mf-ERG) and standard Ganzfeld electroretinography (G-ERG) one week and one month after 1.25 mg intravitreal injections of bevacizumab [57]. Both mf-ERG and G-ERG showed no significant declines in amplitude (>50%) with the mf-ERG patients showing a mean improvement of 35%. One patient had two injections and was followed for three months without significant change from baseline. These small studies suggest that intravitreal bevacizumab has short-term safety with no significant declines in ERG amplitudes after a single injection.

In vitro evaluation of bevacizumab's potential toxicity to human retinal pigment epithelium cells (ARPE19), rat retinal ganglion cells (RGC5) and pig choroidal endothelial cells (CEC) showed no toxicity at the currently injected concentrations, but toxicity was noted at higher levels [58]. The typical concentration of bevacizumab injected into the vitreous is 0.3 mg/ml (1.25 mg in 4 ml) and no toxicity was seen in ARPE19 or RGC5 at concentrations up to 0.8 mg/ml. But after 48 hours at a concentration of 2.5 mg/ml (10.0 mg equivalent intravitreal dose) up to 30% declines in ARPE19 cell number and cell viability was observed. Similarly CEC showed a 65% decline in proliferation rate when exposed to a concentration of 2.5 mg/ml and some moderate amount of DNA synthesis inhibition at lower concentrations of 0.08 mg/ml.

Table 2. Bevacizumab in Age-related Macular Degeneration.

	Number of eyes (n=)	Intravitreal Dose (mg)	Follow-up (months)	ETDRS letter or Snellen VA Change	OCT Change (μm)	Adverse events
Michels S *et al.*	9	5 (mg/kg IV)	3	+12	-177	Mild BP elevation
Rich RM *et al.*	53	1.25	3	+7.9	-100	None
Avery RL *et al.*	81	1.25	3	20/200 to 20/80	-67	None
Spaide RF *et al.*	266	1.25	3	20/184 to 20/109	-90	MI(2), TIA(2), Death (2)*
Bashshur ZF *et al.*	17	2.5	3	20/252 to 20/76 and	-151	None

VA = Visual Acuity, OCT = Optical coherence tomography, IV = Intravenous, BP = Blood Pressure, MI = Myocardial Infarction, TIA = Transient Ischemic Attack

*Not considered directly related to bevacizumab, see text for details

Another *in vitro* report tested short-term safety of human retinal pigment epithelial (ARPE-19), rat neurosensory retinal (R28), and human microvascular endothelial (HMVECad) cells after 2, 6, and 24 hours after exposure to bevacizumab [59]. All cell lines received 0.125 mg/ml, 0.25 mg/ml, 0.50 mg/ml, and 1 mg/ml of bevacizumab (intravitreal concentration 0.3 mg/ml; 1.25 mg/4 ml) with the HMVECad cells being first treated with 5 ng/ml of VEGF and the controls receiving a nonspecific human purified immunoglobulin (IgG). Bevacizumab did not significantly ($p<0.05$) decrease viability of any cell lines at any dose or time point. These two reports show no short-term *in vitro* toxicity of bevacizumab on important retinal cells supporting the results of the previous short-term animal studies.

Age Related Macular Degeneration

The idea of using bevacizumab for ocular disease came from the initial success of another related medication ranibizumab. Clinical experience with ranibizumab and the impressive one year data from the MARINA (Minimally Classic/Occult Trial of the Anti-VEGF Antibody Ranibizumab in the Treatment of Neovascular AMD) trial and the fact that ranibizumab lacked FDA approval motivated Philip Rosenfeld in Miami to use bevacizumab "off-label" in treating neovascular AMD at molecular concentrations similar to those used in the trials evaluating ranibizumab.

Bevacizumab has been used both systemically and via intravitreal injection to treat neovascular AMD. Systemic treatment has been reported in nine patients in the SANA (Systemic Avastin for Neovascular AMD) study [60]. Bilateral disease was present in all but one patient with a unilateral retinal angiomatous proliferation (RAP) with one of the eyes having recent disease progression designated as the study eye. Patients included had visual acuities (VA) 20/40-20/400 and central macular thickness ≥300 μm. Multiple lesions types were studied ("occult," "classic," and RAP), and five of nine study eyes and three of nine fellow eyes had previous photodynamic therapy (PDT). Patients were excluded from the study if a history of uncontrolled hypertension, coagulation abnormalities, previous thromboembolic events (stroke, transient ischemic attacks, myocardial infarction), renal dysfunction, or were treated with anticoagulation therapy, except aspirin, which was allowed.

Initial treatment consisted of three infusions at 5 mg/kg two weeks apart, the same approved dose and interval used for the treatment of metastatic colorectal cancer. The third infusion could be withheld at the discretion of the physician if central thickness decreased by >50% or an adverse event was observed.

Twelve week results for the study eye showed significant mean VA improvement from 20/80 to 20/50 (+12 ETDRS letters) and a 177 μm decrease in central retinal thickness. Similarly, improved VA from 20/160 to 20/80 (+16 ETDRS letters) and reduced central retinal thickness of 92 μm was seen in the fellow eyes. Three lines of VA improvement was observed in 33% (3/9) and 50% (4/8) of study and fellow eyes, respectively.

The only adverse event noted in this study was hypertension. Seven of nine patients required medication to control a significant 12-point mean elevation in systolic blood pressure (no diastolic increases were noted). No thromboembolic or cardiovascular events were identified.

More recently 2.5 mg/kg infusions have been attempted in seven patients with favorable short term results with serum $VEGF_{165}$ levels reduced at twelve weeks after infusion of this lower dose treatment [61]. Even with the use of intravitreal injections, systemic treatment may still be a viable option in certain patients with bilateral disease or if contraindications exist to intravitreal administration.

Rich and Rosenfeld from Bascom Palmer Eye Institute reported a three-month retrospective review of fifty-three eyes of fifty patients who were the first to ever receive intravitreal injections of bevacizumab (1.25 mg in 0.05 ml) in May 2005 [52]. One injection was given at baseline then at the discretion of the physician with 75% of patients having prior treatment with either PDT or pegaptanib. After three months a 7.9 letter (ETDRS) mean gain in vision was observed with 44% gaining 3 lines of vision (15 ETDRS letters) and most patient's visual acuity improvement occurring after the first week. Mean optical coherence tomography (OCT) measurements were 351 μm at baseline with an average improvement of 99.6 μm in central thickness. No thromboembolic events, elevations in blood pressure or other adverse events were noted. Prior treatment with PDT or pegaptanib did not effect the improvement in vision or OCT measurements.

A non-randomized interventional case series with eighty-one eyes in seventy-nine patients with follow-up ranging from 4 to 15 weeks reported similar improvements in vision [62]. Patients were included in this study if they had worsening of the anatomic appearance on FA or OCT with standard treatment of pegaptanib or PDT or if they deferred standard treatment. Rational for exclusion from the study included uncontrolled hypertension or recent myocardial infarction or cerebral vascular accident. The 1.25 mg (0.05 ml) dose was injected intravitrealy every month until the macular edema, subretinal fluid or pigment epithelial detachment resolved. Seventy-eight percent of patients had received prior treatment with PDT (26%), pegaptanib (30%), or a combination of both medicines (22%), and the remaining eyes (22%) bevacizumab was their initial treatment. Mean Snellen visual acuities improved from 20/200 at baseline to 20/125 at four weeks and 20/80 at eight weeks. The average thickness in the central 1 mm of the retina showed significant decrease at one week (61 μm), four weeks (92 μm), eight weeks (89 μm), and twelve weeks (67 μm) as measured on OCT. There were no significant differences in visual acuities or reduction in central thickness seen between the patients previously treated with PDT, pegaptanib and those with bevacizumab as their primary treatment. No adverse events were noted in this study including no elevations in blood pressure.

Spaide and others retrospectively reviewed 266 patients that received in most cases three intravitreal bevacizumab (1.25mg/0.05ml) injections one month apart [63]. Patients initially included were doing poorly on current treatments then expanded to those with large occult or minimally classic lesions with lack of expected benefit from other treatments. Mean visual acuity improved from 20/184 to 20/109 after three months of follow-up and 38.3% showing an improvement and 4.7% a decline in vision. Mean central OCT thickness improved from 340 μm at baseline to 250 μm at one month representing a significant ($p<0.001$) decline of 90 μm. Adverse events noted were: one patient with a non-fatal MI one month after the third injection, one patient had a possible transient ischemic attack who of his own accord stopped his systemic anticoagulation and another with numbness of the lips after injection, with both patients having no residual deficits. Two patients died one with a history of advanced metastatic breast cancer and another patient with a significant history of smoking and emphysema had a MI three months after a single injection. It was discussed that similar adverse events occurred at greater percentages in the larger clinical trials for pegaptanib and photodynamic therapy in both treatment and control groups. They concluded that this number of adverse events is not considered significant given this patient population's age and past medical history.

Another smaller prospective interventional case series of seventeen eyes in seventeen patients found comparable success after three intravitreal injections of bevacizumab at a higher 2.5 mg/0.1 ml dose given four weeks apart [64]. These patients had active neovascular AMD and were not candidates for PDT. After twelve weeks of follow-up the mean VA and central OCT measurements improved from 20/252 to 20/76 and 362 μm to 211 μm, respectively. No systemic or ocular adverse events were noted.

Pathologic Myopia

Intravitreal bevacizumab (1.25 mg) has also been successfully used in patients with choroidal neovascularization resulting from pathologic myopia [65]. A small retrospective review of eleven eyes in nine patients reported an average +3.5 line gain (range, -1 to +8 lines) with a mean follow-up of five months (range, 35 to 224 days). Central foveal thickness improved an average of -103 μm (range, +4 to -356 μm). The treatment benefit was not precluded by previous PDT therapy (5 eyes) and only one injection was required in 8/11 eyes with the remaining three eyes requiring two injections. No complications or drug-related side effects were noted. The results from this small series of patients with pathologic myopia show similar improvement in vision and OCT findings as the previously mentioned reports for bevacizumab in AMD.

Retinal Vascular Disease

Bevacizumab has showed promise in treating proliferative diabetic retinopathy and retinal vascular occlusive disease. Evidence for this was observed in early nonhuman primate studies with iris neovascularization being induced after injection of $VEGF_{165}$ [66] and prevented with injection of an early anti-VEGF monoclonal antibody in eyes with experimentally induced central retinal vein occlusion [67]. Small clinical studies using bevacizumab have found parallel results in humans. Spaide and Fisher reported two patients with extensive vitreous hemorrhage due to proliferative diabetic retinopathy that precluded treatment with panretinal photocoagulation [68]. Both patients had initial intravitreal injections of bevacizumab (1.25 mg in 0.05 ml) and repeat injections based on active neovascularization present on fluorescein angiogram. Rapid improvement of vision coinciding with clearing of the vitreous hemorrhage was observed with VA improving at one week from 20/60 to 20/40 (Patient 1) and from 20/70 to 20/50 (Patient 2). In patient 2 vision improved to 20/25 after a second injection and underwent subsequent panretinal photocoagulation (PRP) for residual neovascularization. Vitrectomy was not required in either patient with Patient 1 having had her vitreous hemorrhage for several years prior to the bevacizumab injection, but no change was seen in the tractional retinal detachments. No adverse events were observed. The authors hypothesized that diabetic vitreous hemorrhages, which were chronic in nature, likely resulted from slow continuous leakage from active vessels and bevacizumab may inhibit the bleeding with the observed clearing of the hemorrhage.

Several cases have been reported using bevacizumab for iris neovascularization (NVI). One patient with proliferative diabetic retinopathy, NVI, an open angle on gonioscopy, normal pressure, and partial PRP 18 months previously, underwent a single injection of bevacizumab (1.25 mg, 0.05 ml) [69]. After one week, the NVI had completely regressed based on slit lamp examination and iris angiography. A second patient presented with VA of 20/70 with florid NVI due to either radiation or diabetes, intraocular pressure (IOP) of 44 mmHg, and twelve clock hours of peripheral anterior synechiae [70]. The patient had previous PRP treatments and cataract extraction several months earlier. Five days after injection, significant regression of iris neovascularization was observed on exam and iris angiography. The patient's IOP was 36 despite maximum medical therapy and underwent trabeculectomy with the angle noted to be uncharacteristically avascular intraoperatively. After approximately two months follow-up the iris neovascularization was absent on slit-lamp exam and iris angiography showed nearly resolved rubeosis with an IOP of seven. The patient's vision improved to 20/30 and no further injections or PRP were given. In a similar report, a patient with neovascular glaucoma who had continued high pressure despite panretinal photocoagulation and cyclophotocoagulation showed a significant reduction in IOP two days after an intravitreal injection 1.0 mg of bevacizumab [71]. Further uses in surgical glaucoma have been reported with a single case of injecting bevacizumab (1.0 mg) into a failed filtering bleb, which resulted in a more diffuse appearing bleb with decreased vascularization [72]. The use of intracameral bevacizumab has been reported in three patients with iris rubeosis associated with glaucoma from proliferative diabetic retinopathy and CRVO [73]. All patients had prior PRP and cyclodestructive procedures prior to intracameral injection of 1.0 mg of bevacizumab in both eyes on separate encounters. Post-injection iris angiography reveled significant decline in rubeosis and leakage within two days after injection that persisted for the four weeks of follow-up. Conclusions were not made regarding the IOP lowering effects of the treatment because of the concomitant treatment of PRP and cyclodestruction. Long-term prospective studies are continuing by the investigators.

Treatment of cystoid macular edema (CME) with bevacizumab secondary to central retinal vein occlusion (CRVO) and cataract extraction has been attempted. The rational for bevacizumab reducing retinal edema is that VEGF induces endothelial cell permeability and nitric oxide production resulting in vascular leakage. *In vitro* studies have confirmed bevacizumab's ability to block these VEGF induced effects [74]. Initially an early single case documented the resolution of CME on OCT after CRVO with improvement of vision from 20/200 to 20/50 one week after injection [75]. A larger study with sixteen eyes in fifteen patients with CRVO reveled similar improvements after intravitreal injections with 1.25 mg/0.05ml of bevacizumab [76]. Fourteen of fifteen patients had previous diagnoses of hypertension and one had diabetes, eleven were on anticoagulation treatment (8 aspirin, 2 warfarin, 1 clopidogrel), nine had previous triamcinolone injections and three had more than five triamcinolone injections. One patient had iris neovascularization and five met criteria as ischemic CRVO according to the CRVO Study Group definition of greater than ten disc areas of nonperfusion. Injections were given at one-month intervals with a mean of 2.8 injections. Visual acuity improved from 20/600 (logMAR = 1.48 ± 0.48) to 20/200 (logMAR = 1.05 ± 0.49) at one month and 20/138 (logMAR = 0.84 ± 0.53) after three months of follow-up and two eyes showed no improvement in vision. Mean arterial venous filling time, defined as the difference between first arterial filling and the end of laminar filling, improved from 62.4 to 20.9 seconds. Central OCT measurements were reduced from an average of 887 μm to 372 μm at one month with previous triamcinolone treatment not affecting the outcomes.

Resolution of cystoid macular edema after cataract surgery was observed in two patients treated with one milligram of intravitreal bevacizumab [77]. Both patients presented with VA of 20/70, CME confirmed with FA and OCT, and a history of treatment failure after five and six months of topical nonsteroidal anti-inflammatory and prednisolone acetate 1%. Complete resolution of CME occurred after ten and fourteen days with improvement in VA to 20/25 and 20/20.

Future Studies with Bevacizumab

The Diabetic Retinopathy Clinical Research Network (DRCR.net) has initiated a Phase II evaluation for diabetic macular edema with intravitreal bevacizumab. Five treatment groups are being evaluated: 1) standard laser photocoagulation, 2) 1.25 mg intravitreal injection at baseline and 6 weeks, 3) 2.5 mg intravitreal intravitreal injection at baseline and 6 weeks, 4) 1.25 mg intravitreal injection at baseline only, 5) 1.25 mg intravitreal injection at baseline, standard laser photocoagulation at 3 weeks, and intravitreal injection at 6 weeks. Results will be evaluated at 12 weeks on completion of the study.

The DRCR.net is also planning to develop a protocol for subjects who have severe non-proliferative or proliferative diabetic retinopathy in whom scatter (panretinal) photocoagulation is planned and who are considered to be at high risk of developing edema within the central subfield or who have macular edema within the central subfield. Subjects would be assigned randomly to one of three groups: 1) no adjunct, 2) intravitreal triamcinolone, or 3) an anti-VEGF medication such as bevacizumab or ranibizumab. The primary outcome would be visual acuity at the three-month exam in order to evaluate morbidity from PRP. This study is still underdevelopment and changes are possible in the exact protocol prior to beginning enrollment.

Cost Analysis and Off-Label Use

The intriguing aspect of bevacizumab and using it "off-label" is the significant cost savings over ranibizumab. With FDA approval and pricing at $1950 per injection ($23,500 per year) for ranibizumab, the cost of bevacizumab ranging between $20-50 per injection ($360 per year) represents a significant savings. Even with patients paying only 20% of the treatment it would still cost $4,700 annual for monthly ranibizumab therapy. For patients with AMD, if feasible, ranibizumab is certainly the safest option having undergone a two year controlled trial. Now that ranibizumab is FDA approved justifying the use of bevacizumab "off-label" will be more difficult. Having two year controlled data with careful safety monitoring for a medicine is far superior to three month data from retrospective reviews given the efficacy appears similar. Because of the significant cost of ranibizumab physicians may find patients unable to afford this treatment. If choosing to use bevacizumab it is important to discuss in detail the risks of "off-label" use of this treatment and the other alternatives. Ophthalmic Mutual Insurance Company (OMIC) offers a informed consent form specifically for intravitreal bevacizumab that may be helpful when discussing this treatment with patients and is available at their website (http://www.omic.com/resources/risk_man/forms.cfm). At this time for patients with AMD ranibizumab is the safest option with far superior evidence to support its use compared to bevacizumab. But with the evidence reported so far there is certainly ample justification for initiating larger controlled trials using bevacizumab and papers continue to be published monthly providing stronger and stronger support for its use.

Conclusion

Anti-VEGF therapies are a new and exciting frontier in treating important vision threatening diseases. The use of monoclonal antibodies or antibody fragments directly against VEGF is just one of a variety of approaches available to decrease VEGF production. Certainly this new paradigm in treatment has provided physicians potent weapons in fighting neovascular ocular diseases and is giving new hope to patients with these previously blinding conditions.

References

[1] Risau W, Flamme I. Vasculogenesis. *Annu Rev Cell Dev Biol* 1995; 11:73-91.

[2] Risau W. Mechanisms of angiogenesis. *Nature* 1997; 386:671-4.

[3] Folkman J. Angiogenesis in cancer, vascular, rheumatoid and other disease. *Nat Med.* 1995; 1:27-31.

[4] Michaelson IC. The mode of development of the vascular system of the retina with some observations on its significance for certain retinal disorders. *Trans Ophthalmol soc Uk* 1948; 68:137-180.

[5] Folkman J. Tumor angiogenesis: therapeutic implications. *N Engl J Med.* 1971; 285:1182-6.

[6] Senger DR, Galli SJ, Dvorak AM, *et al.* Tumor cells secrete a vascular permeability factor that promotes accumulation of ascites fluid. *Science*. 1983; 219:983-5.

[7] Ferrara N, Henzel WJ. Pituitary follicular cells secrete a novel heparin-binding growth factor specific for vascular endothelial cells. *Biochem Biophys Res Commun.* 1989; 16:851-8.

[8] Keck PJ, Hauser SD, Krivi G, *et al.* Vascular permeability factor, an endothelial cell mitogen related to PDGF. *Science.* 1989; 246:1309-12.

[9] Leung DW, Cachianes G, Kuang WJ, *et al.* Vascular endothelial growth factor is a secreted angiogenic mitogen. *Science*. 1989; 246:1306-9.

[10] Olofsson B, Pajusola K, Kaipainen A, *et al.* Vascular endothelial growth factor B, a novel growth factor for endothelial cells. *Proc Natl Acad Sci U S A.* 1996; 93:2576-81.

[11] Joukov V, Pajusola K, Kaipainen A, *et al.* A novel vascular endothelial growth factor, VEGF-C, is a ligand for the Flt4 (VEGFR-3) and KDR (VEGFR-2) receptor tyrosine kinases. *EMBO J.* 1996; 15:1751.
[12] Orlandini M, Marconcini L, Ferruzzi R, *et al.* Identification of a c-fos-induced gene that is related to the platelet-derived growth factor/vascular endothelial growth factor family. *Proc Natl Acad Sci U S A.* 1996; 93:11675-80.
[13] Maglione D, Guerriero V, Viglietto G, *et al.* Isolation of a human placenta cDNA coding for a protein related to the vascular permeability factor. *Proc Natl Acad Sci U S A.* 1991; 88:9267-71.
[14] Stacker SA, Achen MG, Jussila L, *et al.* Lymphangiogenesis and cancer metastasis. *Nat Rev Cancer.* 2002; 2:573-83.
[15] Tischer E, Mitchell R, Hartman T, *et al.* The human gene for vascular endothelial growth factor. Multiple protein forms are encoded through alternative exon splicing. *J Biol Chem.* 1991; 266:11947-54.
[16] Houck KA, Ferrara N, Winer J, *et al.* The vascular endothelial growth factor family: identification of a fourth molecular species and characterization of alternative splicing of RNA. *Mol Endocrinol.* 1991; 5:1806-14.
[17] De Vries C, Escobedo JA, Ueno H, *et al.* The fms-like tyrosine kinase, a receptor for vascular endothelial growth factor. *Science.* 1992; 255:989-91.
[18] Millauer B, Wizigmann-Voos S, Schnurch H, *et al.* High affinity VEGF binding and developmental expression suggest Flk-1 as a major regulator of vasculogenesis and angiogenesis. *Cell.* 1993; 72:835-46.
[19] Jakeman LB, Winer J, Bennett GL, *et al.* Binding sites for vascular endothelial growth factor are localized on endothelial cells in adult rat tissues. *J Clin Invest.* 1992; 89:244-53.
[20] Ferrara N. Vascular endothelial growth factor: basic science and clinical progress. *Endocr Rev* 2004; 25:581-611.
[21] Miller JW, Adamis AP, Shima DT, *et al.* Vascular endothelial growth factor/vascular permeability factor is temporally and spatially correlated with ocular angiogenesis in a primate model. Am J Pathol. 1994; 145:574-84.
[22] Shima DT, Adamis AP, Ferrara N, *et al.* Hypoxic induction of endothelial cell growth factors in retinal cells: identification and characterization of vascular endothelial growth factor (VEGF) as the mitogen. *Mol Med.* 1995; 1:182-93.
[23] Aiello LP, Northrup JM, Keyt BA, *et al.* Hypoxic regulation of vascular endothelial growth factor in retinal cells. *Arch Ophthalmol.* 1995; 113:1538-44.
[24] Kvanta A, Algvere PV, Berglin L, *et al.* Subfoveal fibrovascular membranes in age-related macular degeneration express vascular endothelial growth factor. *Invest Ophthalmol Vis Sci.* 1996; 37:1929-34.
[25] Lopez PF, Sippy BD, Lambert HM, *et al.* Transdifferentiated retinal pigment epithelial cells are immunoreactive for vascular endothelial growth factor in surgically excised age-related macular degeneration-related choroidal neovascular membranes. *Invest Ophthalmol Vis Sci.* 1996; 37:855-68.

[26] Aiello LP, Avery RL, Arrigg PG, *et al.* Vascular endothelial growth factor in ocular fluid of patients with diabetic retinopathy and other retinal disorders. *N Engl J Med.* 1994; 331:1480-7.

[27] Algire GH, Chalkley HW, Legallais FY, *et al.* Vascular reactions of normal and malignant tissues in vivo. I. Vasuclar reactions of mice to wounds and to normal and neoplastic transplants. *J Natl Cancer Inst* 1945, 42:891-899. Mattern J, Koomagi R, Volm M. Coexpression of VEGF and bFGF in human epidermoid lung carcinoma is associated with increased vessel density. *Anticancer Res.* 1997; 17:2249-52.

[28] Volm M, Koomagi R, Mattern J, *et al.* Angiogenic growth factors and their receptors in non-small cell lung carcinomas and their relationships to drug response in vitro. *Anticancer Res.* 1997; 17:99-103.

[29] Volm M, Koomagi R, Mattern J. Prognostic value of vascular endothelial growth factor and its receptor Flt-1 in squamous cell lung cancer. *Int J Cancer.* 1997; 74:64-8.

[30] Brown LF, Berse B, Jackman RW, *et al.* Expression of vascular permeability factor (vascular endothelial growth factor) and its receptors in breast cancer. *Hum Pathol.* 1995; 26:86-91.

[31] Yoshiji H, Gomez DE, Shibuya M, *et al.* Expression of vascular endothelial growth factor, its receptor, and other angiogenic factors in human breast cancer. *Cancer Res.* 1996; 56:2013-6.

[32] Uchida S, Shimada Y, Watanabe G, *et al.* In oesophageal squamous cell carcinoma vascular endothelial growth factor is associated with p53 mutation, advanced stage and poor prognosis. *Br J Cancer.* 1998; 77:1704-9.

[33] Brown LF, Berse B, Jackman RW, *et al.* Expression of vascular permeability factor (vascular endothelial growth factor) and its receptors in adenocarcinomas of the gastrointestinal tract. *Cancer Res.* 1993; 53:4727-35.

[34] Brown LF, Berse B, Jackman RW, *et al.* Increased expression of vascular permeability factor (vascular endothelial growth factor) and its receptors in kidney and bladder carcinomas. *Am J Pathol.* 1993; 143:1255-62.

[35] Olson TA, Mohanraj D, Carson LF, *et al.* Vascular permeability factor gene expression in normal and neoplastic human ovaries. *Cancer Res.* 1994; 54:276-80.

[36] Yamamoto S, Konishi I, Mandai M, *et al.* Expression of vascular endothelial growth factor (VEGF) in epithelial ovarian neoplasms: correlation with clinicopathology and patient survival, and analysis of serum VEGF levels. *Br J Cancer.* 1997; 76:1221-7.

[37] Guidi AJ, Abu-Jawdeh G, Tognazzi K, *et al.* Expression of vascular permeability factor (vascular endothelial growth factor) and its receptors in endometrial carcinoma. *Cancer.* 1996; 78:454-60.

[38] Shweiki D, Itin A, Soffer D, *et al.* Vascular endothelial growth factor induced by hypoxia may mediate hypoxia-initiated angiogenesis. *Nature.* 1992; 359:843-5.

[39] Berkman RA, Merrill MJ, Reinhold WC, *et al.* Expression of the vascular permeability factor/vascular endothelial growth factor gene in central nervous system neoplasms. *J Clin Invest.* 1993; 91:153-9.

[40] Soh EY, Eigelberger MS, Kim KJ, *et al.* Neutralizing vascular endothelial growth factor activity inhibits thyroid cancer growth in vivo. *Surgery.* 2000; 128:1059-65.

[41] Kim KJ, Li B, Winer J, *et al.* Inhibition of vascular endothelial growth factor-induced angiogenesis suppresses tumour growth in vivo. *Nature*. 1993; 362:841-4.

[42] Ferrara N, Hillan KJ, Gerber HP, *et al.* Discovery and development of bevacizumab, an anti-VEGF antibody for treating cancer. *Nat Rev Drug Discov* 2004; 3:391-400.

[43] Yang JC, Haworth L, Sherry RM, *et al.* A randomized trial of bevacizumab, an anti-vascular endothelial growth factor antibody, for metastatic renal cancer. *N Engl J Med.* 2003; 349:427-34.

[44] Miller K, Wang M, Gralow J, et al. A randomized phase III trial of paclitaxel versus paclitaxel plus bevacizumab as first-line therapy for locally recurrent or metastatic breast cancer: a trial coordinated by the Eastern Cooperative Oncology Group (E2100). 28th annual San Antonio Breast Cancer Symposium. San Antonio, Texas. December 2005. Abstract #3.

[45] Sandler AB, Gray R, Bhramer J, et al. Randomized phase II/III Trial of paclitaxel plus carboplatin with or without bevacizumab (NSC # 704865) in patients with advanced non-squamous non-small cell lung cancer (NSCLC): An Eastern Cooperative Oncology Group (ECOG) Trial - E4599. American Society of Clinical Oncology Annual Meeting, May 13, 2005, Abstract LBA4.

[46] Ryan AM, Eppler DB, Hagler KE, *et al.* Preclinical safety evaluation of rhuMAbVEGF, an antiangiogenic humanized monoclonal antibody. *Toxicol Pathol.* 1999; 27:78-86.

[47] Miller KD, Chap LI, Holmes FA, *et al.* Randomized phase III trial of capecitabine compared with bevacizumab plus capecitabine in patients with previously treated metastatic breast cancer. *J Clin Oncol* 2005; 23:792-9.

[48] Kabbinavar FF, Schulz J, McCleod M, *et al.* Addition of bevacizumab to bolus fluorouracil and leucovorin in first-line metastatic colorectal cancer: results of a randomized phase II trial. *J Clin Oncol* 2005; 23:3697-705

[49] Kabbinavar F, Hurwitz HI, Fehrenbacher L, *et al.* Phase II, randomized trial comparing bevacizumab plus fluorouracil (FU)/leucovorin (LV) with FU/LV alone in patients with metastatic colorectal cancer. *J Clin Oncol* 2003; 21:60-5.

[50] Johnson DH, Fehrenbacher L, Novotny WF, *et al.* Randomized phase II trial comparing bevacizumab plus carboplatin and paclitaxel with carboplatin and paclitaxel alone in previously untreated locally advanced or metastatic non-small-cell lung cancer. *J Clin Oncol* 2004; 22:2184-91.

[51] van der Bilt JD, Borel Rinkes IH. Surgery and angiogenesis. *Biochim Biophys Acta* 2004; 1654:95-104.

[52] Rich RM, Rosenfeld PJ, Puliafito CA, *et al.* Short-term safety and efficacy of intravitreal bevacizumab (avastin) for neovascular age-related macular degeneration. *Retina.* 2006; 26:495-511.

[53] Manzano RP, Peyman GA, Khan P, *et al.* Testing intravitreal toxicity of bevacizumab (Avastin). *Retina.* 2006; 26:257-61.

[54] Bakri SJ, Cameron JD, McCannel CA, *et al.* Absence of histologic retinal toxicity of intravitreal bevacizumab in a rabbit model. *Am J Ophthalmol* 2006; 142:162-4.

[55] Shahar J, Avery RL, Heilweil G, *et al.* Electrophysiologic and retinal penetration studies following intravitreal injection of bevacizumab (Avastin). *Retina* 2006; 26:262-9.

[56] Mordenti J, Cuthbertson RA, Ferrara N, *et al.* Comparisons of the intraocular tissue distribution, pharmacokinetics, and safety of 125I-labeled full-length and Fab antibodies in rhesus monkeys following intravitreal administration. *Toxicol Pathol.* 1999; 27:536-44.

[57] Maturi RK, Bleau LA, Wilson DL. Electrophysiologic findings after intravitreal bevacizumab (Avastin) treatment. *Retina* 2006; 26:270-4.

[58] Spitzer MS, Wallenfels-Thilo B, Sierra A, *et al.* Antiproliferative and cytotoxic properties of bevacizumab (avastin) on different ocular cells. *Br J Ophthalmol.* 2006 May 24 (published online).

[59] Luthra S, Narayanan R, Marques LE, *et al.* Evaluation of in vitro effects of bevacizumab (Avastin) on retinal pigment epithelial, neurosensory retinal, and microvascular endothelial cells. *Retina* 2006; 26:512-8.

[60] Michels S, Rosenfeld PJ, Puliafito CA, *et al.* Systemic bevacizumab (Avastin) therapy for neovascular age-related macular degeneration twelve-week results of an uncontrolled open-label clinical study. *Ophthalmology.* 2005; 112:1035-47.

[61] Michels SM, Rosenfeld PJ, Geitzenauer W, *et al.* Standard vs. Reduced Dose Systemic Bevacizumab (Avastin®) for the Treatment of Neovascular Age-Related Macular Degeneration. Association for Research in Vision and Ophthalmology Annual Meeting, Fort Lauderdale, FL. 2006 May 4. Abstract #5231.

[62] Avery RL, Pieramici DJ, Rabena MD, *et al.* Intravitreal bevacizumab (Avastin) for neovascular age-related macular degeneration. *Ophthalmology*. 2006; 113:363-372.

[63] Spaide RF, Laud K, Fine HF, *et al.* Intravitreal bevacizumab treatment of choroidal neovascularization secondary to age-related macular degeneration. *Retina.* 2006; 26:383-90.

[64] Bashshur ZF, Bazarbachi A, Schakal A, *et al.* Intravitreal bevacizumab for the management of choroidal neovascularization in age-related macular degeneration. *Am J Ophthalmol* 2006; 142:1-9.

[65] Yamamoto I, Rogers AH, Reichel E, *et al.* Intravitreal bevacizumab (Avastin) as treatment for subfoveal choroidal neovascularization secondary to pathologic myopia. *Br J Ophthalmol.* 2006 Jul 26; [Epub ahead of print, published online]

[66] Tolentino MJ, Miller JW, Gragoudas ES, *et al.* Vascular endothelial growth factor is sufficient to produce iris neovascularization and neovascular glaucoma in a nonhuman primate. *Arch Ophthalmol.* 1996; 114:964-70.

[67] Adamis AP, Shima DT, Tolentino MJ, *et al.* Inhibition of vascular endothelial growth factor prevents retinal ischemia-associated iris neovascularization in a nonhuman primate. *Arch Ophthalmol.* 1996; 114:66-71.

[68] Spaide RF, Fisher YL. Intravitreal bevacizumab (Avastin) treatment of proliferative diabetic retinopathy complicated by vitreous hemorrhage. *Retina.* 2006; 26:275-8.

[69] Avery RL. Regression of retinal and iris neovascularization after intravitreal bevacizumab (Avastin) treatment. *Retina.* 2006; 26:352-4.

[70] Davidorf FH, Mouser JG, Derick RJ. Rapid improvement of rubeosis iridis from a single bevacizumab (Avastin) injection. *Retina.* 2006; 26:354-6.

[71] Kahook MY, Schuman JS, Noecker RJ. Intravitreal bevacizumab in a patient with neovascular glaucoma. *Ophthalmic Surg Lasers Imaging*. 2006; 37:144-6.

[72] Kahook MY, Schuman JS, Noecker RJ. Needle bleb revision of encapsulated filtering bleb with bevacizumab. *Ophthalmic Surg Lasers Imaging.* 2006; 37:148-50.

[73] Grisanti S, Biester S, Peters S, *et al.*; Tuebingen Bevacizumab Study Group. Intracameral bevacizumab for iris rubeosis. *Am J Ophthalmol.* 2006; 142:158-60.

[74] Wang Y, Fei D, Vanderlaan M, et al. Biological activity of bevacizumab, a humanized antiVEGF antibody in vitro. *Angiogenesis* 2004; 7:335-45.

[75] Rosenfeld PJ, Fung AE, Puliafito CA. Optical coherence tomography findings after an intravitreal injection of bevacizumab (avastin) for macular edema from central retinal vein occlusion. *Ophthalmic Surg Lasers Imaging.* 2005; 36:336-9.

[76] Iturralde D, Spaide RF, Meyerle CB, *et al.* Intravitreal bevacizumab (Avastin) treatment of macular edema in central retinal vein occlusion: a short-term study. *Retina.* 2006; 26:279-84.

[77] Mason JO 3rd, Albert MA Jr, Vail R. Intravitreal bevacizumab (Avastin) for refractory pseudophakic cystoid macular edema. *Retina.* 2006; 26:356-7.

In: New Research on Angiogenesis Inhibitors
Editor: Elmer T. Skinard, pp. 135-146
ISBN: 1-60021-473-8

Chapter VII

ANGIOGENESIS INHIBITORS AND DIABETIC RETINOPATHY

Sho-ichi Yamagishi**, *Kazuo Nakamura, Takanori Matsui, and Takafumi Yoshida
Departments of Internal Medicine, Kurume University School of Medicine, Kurume 830-0011, Japan

ABSTRACT

Diabetic retinopathy is one of the miserable microvascular complications in diabetes and is a leading cause of acquired blindness among the people of occupational age. Chronic hyperglycemia is a major initiator of diabetic retinopathy. Two recent large prospective clinical studies have shown that intensive blood glucose control reduces microvascular complications among patients with diabetes. However, strict control of hyperglycemia is often difficult to maintain and may increase the risk of severe hypoglycemia in diabetic patients. Furthermore, therapeutic options for the treatment of proliferative diabetic retinopathy, photocoagulation and vitrectomy, are limited by considerable side effects. Therefore, to develop novel therapeutic strategies that specifically target diabetic retinopathy is desired for patients with diabetes. This article summarizes the current understanding of anti-angiogenic therapy for diabetic retinopathy, especially focusing on pigment epithelium-derived factor (PEDF).

Keywords: advanced glycation end products, angiogenesis, diabetic retinopathy, pigment epithelium-derived factor

* Correspondence concerning this article should be addressed to Dr. Sho-ichi Yamagishi MD, PhD, Department of Internal Medicine III, Kurume University School of Medicine, Kurume 830-0011, Japan Tel; +81-942-31-7580; Fax; +81-942-31-7707; E-mail; shoichi@med.kurume-u.ac.jp.

Introduction

Diabetic retinopathy is one of the miserable microvascular complications in diabetes and is a leading cause of acquired blindness among the people of occupational age [1]. Chronic hyperglycemia is a major initiator of vascular complications of diabetes. Various hyperglycemia-induced metabolic and hemodynamic derangements, including increased advanced glycation end product (AGE) formation, activation of protein kinase C (PKC), stimulation of the polyol pathway and the renin-angiotensin system (RAS), elicit reactive oxygen species (ROS) generation and subsequently evoke vascular injury, thereby contributing to the characteristic histopathological changes observed in diabetic retinopathy [2]. Two recent large prospective clinical studies, the Diabetes Control and Complications Trial (DCCT) and the United Kigdom Prospective Diabetes Study (UKPDS), have shown that intensive plasma glucose control or blood pressure (BP) lowing effectively reduces the risk of the development and progression of vascular complications among patients with diabetes [3,4]. However, control of hyperglycemia or BP to strict levels is often very difficult to maintain in diabetic patients. Furthermore, therapeutic options for the treatment of proliferative diabetic retinopathy, photocoagulation and vitrectomy, are limited by considerable side effects. Therefore, to develop novel therapeutic strategies that specifically target diabetic retinopathy is desired for patients with diabetes. This article summarizes the current understanding of anti-angiogenic therapy for diabetic retinopathy, especially focusing on pigment epithelium-derived factor (PEDF).

Diabetic Retinopathy

(1) Pathophysiology

The lesions of diabetic retinopathy can be grouped into those associated with background, preproliferative and proliferative retinopathy. The earliest histopathological hallmark of diabetic retinopathy is loss of pericytes [5]. In parallel with loss of pericytes, several characteristic changes including thickening of the basement membrane, hyperpermeability, and formation of microaneurysm are observed [6,7]. These structural and functional abnormalities are followed by microvascular occlusion in the retinas, which ultimately progresses to proliferative changes associated with neovascularization [6,7]. Therefore, it has been postulated that many of these changes are the consequent of the loss of pericytes.

By using pericyte-endothelial cell (EC) co-culture systems, we, along with others have found that pericytes not only regulate the growth, but also preserve the prostacyclin-producing ability and protect against lipid-peroxide-induced injury of ECs, thus playing an important role in the maintenance of microvascular homeostasis [8-10]. Therefore, the loss of pericytes could predispose the vessels to angiogenesis, thrombogenesis, and EC injury, leading to full-blown clinical expression of diabetic retinopathy. Further, recently, Hammes *et al.* investigated the role of capillary coverage with pericytes in early diabetic retinopathy and the contribution to proliferative retinopathy using mice with a single functional allele of platelet-derived growth factor-B (PDGF-B(+/-) mice) [11]. They demonstrated in their studies that retinal capillary coverage with pericytes is crucial for the survival of ECs, particularly under stress conditions such as diabetes, and that pericyte deficiency leads to reduced inhibition of EC proliferation, thus promoting angiogenesis in the retinopathy of prematurity model. Our *in vitro* and their recent *in vivo* observations provide a basis for understanding why diabetic retinopathy develops consequent to pericyte loss, the earliest histopathological hallmarks of diabetic retinopathy.

(2) Angiogenesis Inhibitors

(i) PEDF

PEDF is a glycoprotein that belongs to the superfamily of serine protease inhibitors. It was first purified from the conditioned media of human retinal pigment epithelial cells as a factor with potent neuronal differentiating activity [12]. Recently, PEDF has been shown to be a highly effective inhibitor of angiogenesis in cell culture and animal models. PEDF inhibits the growth and migration of cultured ECs, and it potently suppresses ischemia-induced retinal neovascularization [13,14]. PEDF levels in aqueous humor or vitreous are decreased in diabetic patients, especially with proliferative retinopathy [15-17]. These observations suggest that the loss of PEDF activity in the eye may contribute to the pathogenesis of proliferative diabetic retinopathy.

A non-enzymatic reaction between ketones or aldehydes and the amino groups of proteins contributes to the aging of proteins and to pathological complications of diabetes [18-20]. Under hyperglycemic conditions in diabetes, this process begins with the conversion of reversible Schiff base adducts, and then to more stable, covalently-bound Amadori rearrangement products. Over a course of days to weeks, these early glycation products undergo further reactions and rearrangements to become irreversible crossed-linked, fluorescent protein derivatives termed AGEs. AGEs have been implicated in the development of many of the pathological sequelae of diabetes and aging, such as diabetic retinopathy [21-27]. Furthermore, there is a growing body of evidence that RAGE is a signal-transducing receptor for AGEs and that engagement of RAGE by AGEs evokes inflammatory and proliferative responses in vascular wall cells [21-27]. Indeed, we have recently found that AGEs elicit angiogenesis through the interaction with RAGE [28,29]. It was autocrine vascular endothelial growth factor (VEGF) production in ECs that mainly mediated the angiogenic activity of AGEs. Although the molecular mechanisms of VEGF overexpression elicited by AGEs are not fully understood, our recent investigation has shown that the AGE-RAGE interaction could increase VEGF gene transcription in microvascular ECs by NADPH oxidase-mediated ROS generation and the subsequent nuclear factor-κB (NF-κB) activation via Ras-mitogen activated protein kinase (MAPK) pathway [26,29,30]. PEDF inhibits the AGE-induced angiogenesis and EC activation by suppressing NADPH oxidase activity (unpublished data). Since oxidative stress down-regulates PEDF expression in vascular cells [31-33] and that the levels of vitreous PEDF are decreased in angiogenic eye diseases such as proliferative diabetic retinopathy [15,16], substitution of PEDF may disrupt inappropriate retinal cell responses to AGEs, thus being a promising strategy for treatment of patients with diabetic retinopathy.

Retinal pericytes accumulate AGEs during diabetes, which would be expected to have a detrimental influence on pericyte survival and function [21,22]. AGEs are toxic to retinal pericytes *in vitro*. We have found that AGEs not only induce growth retardation and apoptotic cell death of, but also exert an immediate toxicity to, cultured retinal pericytes [34,35]. Antisense DNA complementary mRNA coding for a receptor for AGEs (RAGE) reversed the AGE-induced decrease in viable cell number of pericytes, while overexpression of RAGE potentiated the deleterious effects of AGEs. These results suggest that the AGE effects on pericytes could be mediated through the interactions with RAGE. Further, we found that ROS generation elicited by the AGE-RAGE interaction mediated these deleterious effects of AGEs [36,37]. Indeed, *N*-acetylcysteine (NAC) or pigment epithelium-derived factor (PEDF), one of the superfamily of serine protease inhibitors with potent neuronal differentiating activity in human retinoblastoma cells, was found to protect cultured retinal pericytes against the AGE-induced apoptosis via its anti-oxidative properties [36,37].

AGEs up-regulated RAGE mRNA levels in pericytes through the intracellular ROS generation as well [35]. These positive feedback loops transduce the AGE signals further, thereby exacerbating the cytopathic effects of AGEs on pericytes. With regard to a source of ROS generation elicited by AGEs, we have recently found that beraprost sodium, a prostacyclin analogue, or forskolin, an activator of adenylate cyclase, protects against the AGE-induced pericyte apoptosis by suppressing ROS generation and subsequent RAGE overexpression [35]. Since cyclic AMP elevating agents were known to block ROS generation in neutrophils by inhibiting NADPH oxidase activity [38], NADPH oxidase might be a source of ROS production elicited by AGEs and also be a target of beraprost sodium or PEDF.

Pericyte dysfunction has also been considered one of the characteristic changes of early phase of diabetic retinopathy. AGEs act on pericytes to stimulate vascular endothelial growth factor (VEGF) expression [35]. VEGF is a specific mitogen to ECs, also known as vascular permeability factor, and is generally thought to be involved in the pathogenesis of proliferative diabetic retinopathy. Indeed, some clinical observations have demonstrated that VEGF level in ocular fluid is positively correlated with the activity of neovascularization in diabetic retinopathy [39,40]. Recently, VEGF level was also found to be associated with the breakdown of the blood-retinal barrier, thus being involved in microvascular hyperpermeability in background retinopathy [41-43]. These observations suggest that AGE-RAGE-elicited ROS generation might be involved in the development of early diabetic retinopathy by inducing VEGF expression in pericytes as well.

Angiopoietin (ang)-Tie receptor interaction plays an important role in both physiological and pathological angiogenesis [44]. Engagement of tie-2 by ang-1 has been known to promote recruitment of pericytes, thereby supporting the establishment and maintenance of vascular integrity; while ang-2 is a naturally occurring antagonist of ang-1, and induces the loosening of contacts between ECs and pericytes [44-46]. AGEs and H_2O_2 increase a ratio of ang-2 to ang-1 mRNA level and simultaneously up-regulate VEGF mRNA levels in microvascular ECs and pericytes, respectively [29,47]. A plastic window for blood vessel remodeling is defined by pericyte coverage of the preformed endothelial network [48].

Therefore, the AGE-elicited oxidative stress-induced pericyte apoptosis and increased ratio of ang-2/ang-1 could disrupt the pericyte-EC interactions, thus promoting angiogeneis by acting in concert with VEGF. Further, we have recently found that PEDF completely reversed an increased mRNA ratio of ang-2 to ang-1 and an up-regulation of VEGF mRNA levels in H_2O_2-exposed pericytes [47]. These results suggest that PEDF might prevent the alteration of angiogenesis-related gene expression in pericytes through its anti-oxidative ability.

There has been increasing interest in the role of inflammatory reaction in diabetic retinopathy [49]. AGEs are implicated in the process of vascular inflammation as well. Indeed, AGEs have been recently shown to increase leukocyte adhesion to cultured retinal microvascular ECs by inducing intracellular cell adhesion molecule-1 (ICAM-1) expression [50]. This phenomenon is also apparent in non-diabetic mice infused with preformed AGEs, which results in significant leukostasis and blood-retinal barrier dysfunction in these mice [50]. Retinal VEGF has been reported to induce ICAM-1 expression, thus leading to leukostasis and breakdown of blood-retinal barrier *in vivo* [51-53]. Recently, we have found that administration of PEDF or pyridoxal phosphate, an AGE inhibitor, decreases retinal levels of 8-hydroxydeoxyguanosine (8-OHdG), an oxidative stress marker, and subsequently suppresses ICAM-1 gene expression and retinal leukostasis in diabetic rats [54]. Further, intravenous administration of AGE to normal rats increases ICAM-1 gene expression and retinal leukostasis, which are blocked by PEDF. PEDF also inhibites the AGE-induced T cell adhesion to microvascular endothelial cells by suppressing ICAM-1 expression. These observations demonstrate that PEDF inhibites diabetes- or AGE-elicited retinal leukostasis by suppressing ICAM-1. Further, PEDF also blocks monocyte chemoattractant protein-1 (MCP-1) expression in AGE-exposed microvascular ECs through its anti-oxidative properties [55]. Since the levels of MCP-1 in vitreous fluids are correlated with the severity of proliferative diabetic retinopathy [56], AGEs would be one of the key pro-inflammatory factors for progression of diabetic retinopathy. Taken together, PEDF may play a protective role against early phase of diabetic retinopathy by attenuating the deleterious effect of AGE.

The local rennin-angiotensin system (RAS) is activated under diabetes [57]. We have very recently found that PEDF inhibited the angiotensin II (Ang II)-induced ICAM-1 mRNA up-regulation in microvascular ECs [58]. In addition, MOLT-3 T cell adhesion to ECs was also blocked by the treatment of PEDF. Since there is increasing evidence to show the active involvement of the RAS in early stage of diabetic retinopathy [59-61], PEDF could play a protective role against early diabetic retinopathy by suppressing RAS-mediated ICAM-1-elicited leukostasis in diabetic retinas.

(ii) Endostatin

Endostatin is an endogenous collagen XVIII–derived angiogenesis inhibitor identified and purified from a murine hemangioendothelioma cell line [62] and later characterized in mice [63]. It corresponds to a 20-kDa fragment derived from the COOH-terminal NC1 domain of type XVIII collagen [62,63]. Endostatin binds to heparin and with low affinity to all surface heparan sulfate proteoglycans that are involved in growth factor signaling [64]. The anti-angiogenic activity of endostatin seems to depend on the interactions with heparan sulfate proteoglycans, possibly by an interaction between discontinuous sulfated domains in heparan sulfate proteoglycans and arginine clusters on the surface of endostatin [65]. Further, endostatin also inhibits activation of certain matrix metalloproteinases (MMP; i.e., MMP-2, -9, and -13 and MT1-MMP) and binds directly to MMP-2 and -9 [66], thereby blocking angiogenesis *in vivo*. In addition to MMPs, endostatin has been shown to interfere with the action of other proteases, like the plasminogen activator system [67].

Recently, intraocular expression of endostatin has been found to reduce VEGF-induced retinal vascular permeability, neovascularization, and retinal detachment [68]. These observations suggest that endostatin may be an endogenous inhibitor of vasopermeability as well as neovascularization. In patients with diabetic retinopathy, endostatin gene transfer may provide a way to decrease the risk of three causes of visual loss: macular edema, neovascularization, and retinal detachment.

(iii) Angiostatin

Angiostatin, a naturally occurring inhibitor of angiogenesis, was discovered based on its ability to block tumor growth *in vivo* by inhibiting the formation of new tumor blood vessels [69,70]. Angiostatin is a proteolytically derived internal fragment of plasminogen and may contain various members of the 5 plasminogen "kringle" domains, depending on the exact sites of proteolysis. Binding of angiostatin to the alpha/beta-subunits of ATP synthase on the cell surface may mediate its antiangiogenic effects and the down-regulation of endothelial cell proliferation and migration [71]. Recently, recombinant angiostatin has been found to prevent retinal neovascularization in a murine proliferative retinopathy model [72]. The findings suggest that angiostatin could have therapeutic potential for neovascular retinopathies such as proliferative diabetic retinopathy. Moreover, plasminogen kringle 5 was recently reported to inhibit ischemia-induced retinal neovascularization in a rat model by down-regulating VEGF and up-regulating PEDF [73]. The observations suggest that the anti-angiogenic and growth inhibitory effects of angiostatin on diabetic retinopathy could mediated, at least in part, through PEDF activity.

Conclusion

There is a growing body of evidence that angiogenesis inhibitors could play a protective role against the pathogenesis of diabetic retinopathy. Pharmacological up-regulation or substitution of angiogenesis inhibitors such as PEDF may offer a promising strategy for halting the development and progression of diabetic retinopathy.

Acknowledgement

This study was supported in part by Grants of Venture Research and Development Centers from the Ministry of Education, Cultures, Sports, Science and Technology, Japan (S. Y.).

REFERENCES

[1] Frank RN. On the pathogenesis of diabetic retinopathy. A 1990 update. *Ophthalmology* 199; 98: 586-93.

[2] Brownlee M. Biochemistry and molecular cell biology of diabetic complications. *Nature* 2001; 414: 813-20.

[3] The Diabetes Control and Complications Trial Research Group. The effect of intensive treatment of diabetes on the development and progression of long-term complications in insulin-dependent diabetes mellitus. *N Engl J Med* 1993; 329: 977-86.

[4] UK Prospective Diabetes Study (UKPDS) Group. Intensive blood-glucose control with sulphonylureas or insulin compared with conventional treatment and risk complications in patients with type 2 diabetes (UKPS 33). *Lancet* 1998; 352: 837-53.

[5] Cogan DG, Toussaint D, Kuwabara T. Retinal vascular patterns. IV. Diabetic retinopathy. *Arch Ophthalmo* 1961; 66: 366-78.

[6] Sims DE. Recent advances in pericyte biology--implications for health and disease. *Can J Cardiol* 1991; 7: 431-43.

[7] Yamagishi S, Imaizumi T. Pericyte biology and diseases. *Int J Tissue React* 2005; 27: 125-35.

[8] Dodge AB, D'Amore PA. Cell-cell interactions in diabetic angiopathy. *Diabetes Care* 1992; 15: 1168-80.

[9] Yamagishi S, Kobayashi K, Yamamoto H. Vascular pericytes not only regulate growth, but also preserve prostacyclin-producing ability and protect against lipid peroxide-induced injury of co-cultured endothelial cells. *Biochem Biophys Res Commun* 1993; 190: 418-25.

[10] Yamagishi S, Hsu CC, Kobayashi K, Yamamoto H. Endothelin 1 mediates endothelial cell-dependent proliferation of vascular pericytes. *Biochem Biophys Res Commun* 1993; 191: 840-6.

[11] Hammes HP, Lin J, Renner O, Shani M, Lundqvist A, et al. Pericytes and the pathogenesis of diabetic retinopathy. *Diabetes* 2002; 51: 3107-12.

[12] Tombran-Tink J, Chader CG, Johnson LV. PEDF: Pigment epithelium-derived factor with potent neuronal differentiative activity. *Exp Eye Res* 1991; 53: 411-4.

[13] Dawson DW, Volpert OV, Gillis P, Crawford SE, Xu HJ, Benedict W, Bouck NP. Pigment epithelium-derived factor: A potent inhibitor of angiogenesis. *Science* 1999; 285: 245-8.

[14] Duh EJ, Yang HS, Suzuma I, Miyagi M, Youngman E, Mori K, Katai M, Yan L, Suzuma K, West K, Davarya S, Tong P, Gehlbach P, Pearlman J, Crabb JW, Aiello LP, Campochiaro PA, Zack DJ. Pigment epithelium-derived factor suppresses ischemia-induced retinal neovascularization and VEGF-induced migration and growth. *Invest Ophthalmol Vis Sci* 2002; 43: 821-9.

[15] Spranger J, Osterhoff M, Reimann M, Mohlig M, Ristow M, Francis MK, Cristofalo V, Hammes HP, Smith G, Boulton M, Pfeiffer FH. Loss of the antiangiogenic pigment epithelium-derived factor in patients with angiogenic eye disease. *Diabetes* 2002; 50: 2641-5.

[16] Ogata N, Tombran-Tink J, Nishikawa M, Nishimura T, Mitsuma Y, Sakamoto T, Matsumura M. Pigment epithelium-derived factor in the vitreous is low in diabetic retinopathy and high in rhegmatogenous retinal detachment. *Am J Ophthalmol* 2001; 132: 378-82.

[17] Boehm BO, Lang G, Volpert O, jehle PM, Kurkhaus A, Rosinger S, Lang GK, Bouck N. Low content of the natural ocular anti-angiogenic agent pigment epithelium-derived factor (PEDF) in aqueous humor predicts progression of diabetic retinopathy. *Diabetologia* 2003; 46: 394-400.

[18] Brownlee M, Cerami A, Vlassara H. Advanced glycosylation end products in tissue and the biochemical basis of diabetic complications. *N Engl J Med* 1988; 318: 1315-21.

[19] Dyer DG, Blackledge JA, Thorpe SR, Baynes JW. Formation of pentosidine during nonenzymatic browning of proteins by glucose. Identification of glucose and other carbohydrates as possible precursors of pentosidine in vivo. *J Biol Chem* 1991; 266: 11654-60.

[20] Grandhee SK, Monnier VM. Mechanism of formation of the Maillard protein cross-link pentosidine. Glucose, fructose, and ascorbate as pentosidine precursors. *J Biol Chem* 1991; 266: 11649-53.

[21] Stitt AW. The role of advanced glycation in the pathogenesis of diabetic retinopathy. *Experimental and Molecular Pathology* 2003; 75: 95-108.

[22] Stitt AW, Li YM, Gardiner TA, Bucala R, Archer DB, et al. Advanced glycation end products (AGEs) co-locarize with AGE receptors in the retinal vasculature of diabetic and AGE-infused rats. *Am J Pathol* 1997; 150: 523-31.

[23] Yamagishi S, Nakamura K, Jinnouchi Y, Takenaka K, Imaizumi T. Therapeutic implications of blockers of advanced glycation end products (AGEs)-their receptor (RAGE) system. *Int J Pharm* 2005; 2: 203-9.

[24] Yamagishi S, Takeuchi M, Inagaki Y, Nakamura K, Imaizumi T. Role of advanced glycation end products and their receptor (RAGE) in the pathogenesis of diabetic microangiopathy. *Int J Clin Pharm Res* 2003; 23: 129-34.

[25] Takeuchi M, Yamagishi S, Iwaki M, Nakamura K, Imaizumi T. Advanced glycation end product (AGE) inhibitors and their therapeutic implications in diseases. *Int J Clin Pharm Res* 2004; 24: 95-101.

[26] Yamagishi S, Imaizumi T. Diabetic vascular complications: pathophysiology, biochemical basis and potential therapeutic strategy. *Curr Pharm Des* 2005; 11: 2279-99.

[27] Yamagishi S, Nakamura K, Imaizumi T. Advanced glycation end products (AGEs) and diabetic vascular complications. *Curr Diabetes Rev* 2005; 1: 93-106.

[28] Yamagishi S, Yonekura H, Yamamoto Y, Katsuno K, Sato F, et al. Advanced glycation end products-driven angiogenesis in vitro. Induction of the growth and tube formation of human microvascular endothelial cells through autocrine vascular endothelial growth factor. *J Biol Chem* 1997; 272: 8723-30.

[29] Okamoto T, Yamagishi S, Inagaki Y, Amano S, Koga K, et al. Angiogenesis induced by advanced glycation end products and its prevention by cerivastatin. *FASEB J* 2002; 16: 1928-30.

[30] Okamoto T, Yamagishi S, Inagaki Y, Amano S, Takeuchi M, et al. Incadronate disodium inhibits advanced glycation end products-induced angiogenesis in vitro. *Biochem Biophys Res Commun* 2002; 297: 419-24.
[31] Yamagishi S, Nakamura K, Ueda S, Kato S, Imaizumi T. Pigment epithelium-derived factor (PEDF) blocks angiotensin II signaling in endothelial cells via suppression of NADPH oxidase: a novel anti-oxidative mechanism of PEDF. *Cell Tissue Res* 2005; 320: 437-45.
[32] Yamagishi S, Inagaki Y, Nakamura K, Abe R, Shimizu T, et al. Pigment epithelium-derived factor (PEDF) inhibits TNF-α-induced interleukin-6 expression in endothelial cells by suppressing NADPH oxidase-mediated reactive oxygen species generation. *J Mol Cell Cardiol* 2004; 37: 497-506.
[33] Yamagishi S, Inagkai Y, Amano S, Okamoto T, Takeuchi M. Up-regulation of vascular endothelial growth factor and down-regulation of pigment epithelium-derived factor mRNA in leptin-exposed cultured retinal pericytes. *Int J Tissue React* 2002; 24: 137-142.
[34] Yamagishi S, Hsu CC, Taniguchi M, Harada S, Yamamoto Y, et al. Receptor-mediated toxicity to pericytes of advanced glycosylation end products: a possible mechanism of pericyte loss in diabetic microangiopathy. *Biochem Biophys Res Commun* 1995; 213: 681-7.
[35] Yamagishi S, Amano S, Inagaki Y, Okamoto T, Koga K, et al. Advanced glycation end products-induced apoptosis and overexpression of vascular endothelial growth factor in bovine retinal pericytes. *Biochem Biophys Res Commun* 2002; 290: 973-8.
[36] Yamagishi S, Inagaki Y, Amano S, Okamoto T, Takeuchi M, et al. Pigment epithelium-derived factor protects cultured retinal pericytes from advanced glycation end product-induced injury through its antioxidative properties. *Biochem Biophys Res Commun* 2002; 296: 877-82.
[37] Yamagishi S, Amano S, Inagaki Y, Okamoto T, Takeuchi M, et al. Beraprost sodium, a prostaglandin I_2 analogue, protects against advanced gycation end products-induced injury in cultured retinal pericytes. *Mol Med* 2002; 8: 546-50.
[38] Ottonello L, Morone MP, Dapino P, Dallegri F. Tumor necrosis factor-alpha-induced oxidative burst in neutrophils adherent to fibronectin: effects of cyclic AMP-elevating agents. *Br J Haematol* 1995; 91: 566-70.
[39] Adamis AP, Miller JW, Bernal MT, D'Amico DJ, Folkman J, et al. Increased vascular endothelial growth factor levels in the vitreous of eyes with proliferative diabetic retinopathy. *Am J Ophthalmol* 1994; 118: 445-50.
[40] Aiello LP, Avery RL, Arrigg PG, Keyt BA, Jampel HD, et al. Vascular endothelial growth factor in ocular fluid of patients with diabetic retinopathy and other retinal disorders. *N Engl J Med* 1994; 331: 1480-7.
[41] Murata T, Ishibashi T, Khalil A, Hata Y, Yoshikawa H, et al. Vascular endothelial growth factor plays a role in hyperpermeability of diabetic retinal vessels. *Ophthalmic Res* 1995; 27: 48-52.
[42] Stitt AW, Bhaduri T, McMullen CB, Gardiner TA, Archer DB. Advanced glycation end products induce blood-retinal barrier dysfunction in normoglycemic rats. *Mol Cell Biol Res Commun* 2000; 3: 380-8.

[43] Segawa Y, Shirao Y, Yamagishi S, Higashide T, Kobayashi M, et al. Upregulation of retinal vascular endothelial growth factor mRNAs in spontaneously diabetic rats without ophthalmoscopic retinopathy. A possible participation of advanced glycation end products in the development of the early phase of diabetic retinopathy. *Ophthalmic Res* 1998; 30: 333-9.

[44] Folkman J. D'Amore PA. Blood Vessel Formation: What Is Its Molecular Basis? *Cell* 1996; 87: 1153-5.

[45] Maisonpierre PC, Suri C, Jones PF, Bartunkova S, Wiegand SJ, et al. Angiopoietin-2, a Natural Antagonist for Tie2 That Disrupts in vivo Angiogenesis. *Science* 1997; 277: 55-60.

[46] Jonnes N, Iljin K, Dumont DJ, Alitalo K. Tie receptors: new modulators of angiogenic and lymphangiogenic responses. *Nat Rev Mol Cell Biol* 2001; 2: 257-67.

[47] Amano S, Yamagishi S, Inagaki Y, Nakamura K, Takeuchi M, et al. Pigment epithelium-derived factor inhibits oxidative stress-induced apoptosis and dysfunction of cultured retinal pericytes. *Microvas Res* 2005; 69: 45-55.

[48] Benjamin LE, Hemo I, Keshet E. A plasticity window for blood vessel remodelling is defined by pericyte coverage of the preformed endothelial network and is regulated by PDGF-B and VEGF. *Development* 1998; 125: 1591-8.

[49] Schroder S, Palinski W, Schmid-Schonbein GW. Activated monocytes and granulocytes, capillary nonperfusion, and neovascularization in diabetic retinopathy. *Am J Pathol* 1991; 139: 81-100.

[50] Moore TC, Moore JE, Kaji Y, Frizzell N, Usui T, et al. The role of advanced glycation end products in retinal microvascular leukostasis. *Invest Ophthalmol Vis Sci* 2003; 44: 4457-64.

[51] Lu M, Perez VL, Ma N, Miyamoto K, Peng HB, et al. VEGF increases retinal vascular ICAM-1 expression in vivo. *Invest Ophthalmol Vis Sci* 1999; 40: 1808-12.

[52] Joussen AM, Poulaki V, Qin W, Kirchhof B, Mitsiades N, et al. Retinal vascular endothelial growth factor induces intercellular adhesion molecule-1 and endothelial nitric oxide synthase expression and initiates early diabetic retinal leukocyte adhesion in vivo. *Am J Pathol* 2002; 160: 501-9.

[53] Ishida S, Usui T, Yamashiro K, Kaji Y, Ahmed E, et al. VEGF164 is proinflammatory in the diabetic retina. *Invest Ophthalmol Vis Sci* 2003; 44: 2155-62.

[54] Yamagishi S, Matsui T, Nakamura K, Takeuchi M, Imaizumi T. Pigment epithelium-derived factor (PEDF) prevents diabetes- or advanced glycation end products-elicited retinal leukostasis. *Microvas Res* 2006 (in press).

[55] Inagaki Y, Yamagishi S, Okamoto T, Takeuchi M, Amano S. Pigment epithelium-derived factor prevents advanced glycation end products-induced monocyte chemoattractant protein-1 production in microvascular endothelial cells by suppressing intracellular reactive oxygen species generation. *Diabetologia* 2003; 46: 284-7.

[56] Mitamura Y, Takeuchi S, Matsuda A, Tagawa Y, Mizue Y, et al. Monocyte chemotactic protein-1 in the vitreous of patients with proliferative diabetic retinopathy. *Ophthalmologica* 2000; 215: 415-8.

[57] Anderson S. Role of local and systemic angiotensin in diabetic renal disease. *Kidney Int Suppl* 1997; 63: S107-10.

[58] Yamagishi S, Kikuchi S, Nakamura K, Matsui T, Makino T, et al. Pigment epithelium-derived factor (PEDF) blocks angiotensin II-induced T cell adhesion to endothelial cells by suppressing intercellular adhesion molecule-1. *Hor Metab Res* 2006 (in press).
[59] Nagisa Y, Shintani A, Nakagawa S. The angiotensin II receptor antagonist candesartan cilexetil (TCV-116) ameliorates retinal disorders in rats. *Diabetologia* 2001; 44: 883-8.
[60] Gilbert RE, Kelly DJ, Cox AJ, Wilkinson-Berka JL, Rumble JR, et al. Angiotensin converting enzyme inhibition reduces retinal overexpression of vascular endothelial growth factor and hyperpermeability in experimental diabetes. *Diabetologia* 2000; 43: 1360-7.
[61] Moravski CJ, Skinner SL, Stubbs AJ, Sarlos S, Kelly DJ, et al. The renin-angiotensin system influences ocular endothelial cell proliferation in diabetes: transgenic and interventional studies. *Am J Pathol* 2003; 162: 151-60.
[62] O'Reilly MS, Boehm T, Shing Y, Fukai N, Vasios G, et al Endostatin: an endogenous inhibitor of angiogenesis and tumor growth. *Cell* 1997; 88: 277-85.
[63] Standker L, Schrader M, Kanse SM, Jurgens M, Forssmann WG, Preissner KT. Isolation and characterization of the circulating form of human endostatin. *FEBS Lett* 1997; 420: 129-33.
[64] Karumanchi SA, Jha V, Ramchandran R, Karihaloo A, Tsiokas L, et al. Cell surface glypicans are low-affinity endostatin receptors. *Mol Cell* 2001; 7: 811-22.
[65] Kreuger J, Matsumoto T, Vanwildemeersch M, Sasaki T, Timpl R, et al. Role of heparan sulfate domain organization in endostatin inhibition of endothelial cell function. *EMBO J* 2002; 21: 6303-11.
[66] Nyberg P, Heikkila P, Sorsa T, Luostarinen J, Heljasvaara R, et al. Endostatin inhibits human tongue carcinoma cell invasion and intravasation and blocks the activation of matrix metalloprotease-2, -9, and -13. *J Biol Chem* 2003; 278: 22404-11.
[67] Wickstrom SA, Veikkola T, Rehn M, Pihlajaniemi T, Alitalo K, Keski-Oja J. Endostatin-induced modulation of plasminogen activation with concomitant loss of focal adhesions and actin stress fibers in cultured human endothelial cells. *Cancer Res* 2001; 61: 6511-6.
[68] Takahashi K, Saishin Y, Saishin Y, Silva RL, Oshima Y, et al. Intraocular expression of endostatin reduces VEGF-induced retinal vascular permeability, neovascularization, and retinal detachment. *FASEB J* 2003; 17: 896-8.
[69] O'Reilly. Angiostatin: an endogenous inhibitor of angiogenesis and of tumor growth. *EXS* 1997; 79: 273-94.
[70] Wahl ML, Kenan DJ, Gonzalez-Gronow M, Pizzo SV. Angiostatin's molecular mechanism: aspects of specificity and regulation elucidated. *J cell Biochem* 2005; 96: 242-61.
[71] Moser TL, Stack MS, Asplin I, Enghild JJ, Hojrup P, et al. Angiostatin binds ATP synthase on the surface of human endothelial cells. *Proc Natl Acad Sci USA* 1999; 96: 2811-6.
[72] Meneses PI, Hajjar KA, Berns KI, Duvoisin RM. Recombinant angiostatin prevents retinal neovascularization in a murine proliferative retinopathy model. *Gene Ther* 2001; 8: 646-8.

[73] Gao, G, Li Y, Gee S, Dudley A, Fant J, et al. Down-regulation of vascular endothelial growth factor and up-regulation of pigment epithelium-derived factor: a possible mechanism for the anti-angiogenic activity of plasminogen kringle 5. *J Biol Chem* 2002; 277: 9492-7.

INDEX

A

accessibility, 92
accounting, vii, 1
accumulation, 31, 41, 81, 82, 88, 89, 90, 92, 93, 128
acetic acid, 82
acetone, 108
acid, 6, 27, 46
activation, 3, 5, 6, 13, 16, 17, 33, 34, 35, 48, 50, 99, 106, 107, 111, 112, 136, 137, 140, 145
acute lymphoblastic leukemia, 64
adalimumab, 26
adenocarcinoma, 35, 72, 107
adenovirus, 8, 19
adhesion, viii, 5, 13, 19, 23, 25, 26, 42, 103, 110, 112, 139, 144, 145
adhesions, 145
adjunctive therapy, 118
adults, vii, ix, 1, 29, 55, 57
adverse event, xi, 116, 119, 121, 123, 124, 125
aetiology, 25
affect, x, 7, 13, 30, 57, 62, 67, 81, 83, 85, 88, 94, 99, 103
age, x, xi, 115, 116, 124, 129, 131, 132, 135, 136
agent, ix, 7, 8, 13, 14, 19, 20, 25, 55, 56, 61, 62, 63, 64, 68, 94, 100, 104, 106, 107, 113, 142
aggregates, 25
aggregation, 31
aging, 137
agonist, 65
aldehydes, 137
alkaloids, ix, 56, 64
alkylation, 100
allele, 52, 136
alpha-fetoprotein, 8
alternative(s), 13, 27, 31, 37, 128, 129
amino acids, 27, 117
ammonium, 82
amplitude, 121, 122
amyotrophic lateral sclerosis, 37
androgen, 101, 109
angiogenesis, vii, viii, ix, 2, 3, 4, 5, 6, 7, 11, 12, 13, 14, 15, 16, 17, 18, 19, 20, 21, 24, 26, 27, 28, 29, 30, 31, 32, 33, 34, 35, 36, 37, 38, 39, 40, 42, 43, 45, 48, 49, 50, 51, 52, 53, 54, 55, 56, 59, 60, 61, 62, 65, 67, 68, 69, 70, 71, 72, 73, 74, 75, 79, 93, 97, 98, 100, 101, 103, 105, 107, 109, 110, 111, 112, 113, 116, 117, 119, 128, 129, 130, 131, 135, 136, 137, 138, 139, 140, 141, 142, 143, 145
angiogenic process, 4, 6, 13, 15
angiogram, 125
angiography, 126
angiotensin II, 95, 139, 143, 145
angiotensin II receptor antagonist, 145
animals, 7, 82, 83, 91
ankylosing spondylitis, 28, 41
antiangiogenic agents, x, 9, 12, 14, 68, 71, 77, 79, 93, 108
antibiotic, 38, 118
antibody, viii, x, 12, 20, 24, 26, 27, 28, 29, 36, 37, 41, 63, 71, 78, 80, 82, 87, 94, 95, 105, 115, 118, 121, 128, 131, 133
anti-cancer, 19, 111, 118
anticancer drug, 104
anticoagulation, 120, 123, 124, 126
antigen, 65, 78, 80, 95
anti-inflammatory agents, 120
antimetabolites, ix, 56
antioxidant, 100
antisense, 17
anti-TNFα, viii, 23, 24, 26, 28

antitumor, ix, x, 17, 19, 56, 59, 60, 62, 63, 65, 72, 74, 97, 99, 100, 102, 104, 109, 111, 112
antitumor agent, 100
aorta, 109
apoptosis, x, 2, 5, 6, 10, 11, 18, 19, 28, 38, 40, 54, 60, 61, 62, 69, 70, 71, 99, 101, 105, 106, 107, 109, 111, 138, 139, 143, 144
aqueous humor, 137, 142
arginine, 139
arrest, 7, 11, 20, 105, 113
arteries, 4
arthritis, viii, 24, 25, 26, 30, 31, 36, 37, 38, 39, 40, 44, 45, 47, 51, 52, 53, 54
articular cartilage, 25, 32, 46
ascites, 80, 90, 91, 128
ascorbic acid, 100
assessment, 12, 13, 59, 62, 73
association, 61, 63, 106, 119
atherosclerosis, 39, 51
ATP, 102, 140, 145
autopsy, 91

B

bacteria, 100
barriers, 79, 97
basement membrane, 2, 4, 19, 26, 33, 34, 117, 136
benzene, 99
bilirubin, 11
binding, 6, 17, 21, 27, 29, 30, 31, 34, 36, 44, 47, 49, 50, 60, 78, 80, 93, 97, 102, 103, 104, 105, 108, 117, 128, 129
biological activity, 104
biomarkers, 13
biopsy, 46
biosynthesis, 6
bladder, 19, 70, 117, 130
bleeding, 11, 119, 120, 125
blindness, xi, 135, 136
blocks, 21, 34, 60, 102, 139, 143, 145
blood, vii, viii, ix, x, xi, 1, 2, 4, 6, 12, 24, 25, 26, 28, 29, 30, 31, 34, 36, 37, 38, 42, 52, 56, 59, 61, 63, 68, 69, 79, 82, 98, 99, 100, 103, 105, 109, 115, 116, 117, 118, 124, 135, 136, 138, 139, 140, 141, 143, 144
blood flow, vii, ix, 6, 56, 68, 98
blood pressure, 124, 136
blood supply, 61, 79, 117
blood vessels, vii, viii, 1, 2, 4, 24, 26, 28, 29, 30, 31, 34, 36, 38, 59, 63, 100, 103, 109, 116, 118, 140
body fluid, 33
body weight, 82, 88, 94
bone growth, 34, 36
bone marrow, x, 4, 13, 15, 29, 31, 35, 43, 57, 60, 61, 63, 66, 67, 77, 79, 82, 83, 94
bone marrow transplant, 94
brain, 60, 70, 73
brain tumor, 60, 70, 73
branching, 30, 51, 105
breakdown, 32, 33, 34, 35, 138, 139
breast cancer, 17, 34, 35, 37, 59, 62, 64, 65, 66, 68, 71, 72, 73, 74, 75, 76, 79, 95, 97, 104, 108, 109, 118, 124, 130, 131
breast carcinoma, 50, 97

C

calcification, 112
California, 1
cancer, v, vii, viii, ix, x, 1, 2, 5, 6, 8, 11, 12, 13, 15, 16, 18, 19, 20, 24, 27, 35, 37, 38, 39, 52, 55, 56, 57, 59, 60, 63, 64, 66, 68, 69, 70, 71, 72, 73, 74, 75, 77, 78, 79, 91, 92, 94, 95, 96, 97, 101, 102, 103, 106, 108, 109, 111, 112, 113, 115, 116, 117, 118, 128, 129, 131
cancer cells, 2, 35, 38, 39, 77, 91, 102, 103, 108, 113
cancer progression, 5
candidates, 125
cannabinoids, 109, 110
capillary, 27, 30, 34, 35, 74, 100, 107, 117, 136, 144
carbonyl groups, 108
carcinogenesis, 4, 15, 20, 49, 113
carcinoma, v, vii, 1, 3, 10, 14, 19, 27, 35, 51, 62, 73, 74, 95, 106, 108, 111, 117, 118, 130, 145
cardiovascular disease, 25, 119, 120
carrier, 100
cartilage, 25, 26, 32, 33, 34, 35, 36, 46, 52
caspases, 106, 111
cataract, 126, 127
cataract extraction, 126
CD95, 60, 69
cDNA, 129
CEC, 122
cell, viii, x, 2, 4, 5, 6, 7, 10, 11, 12, 13, 16, 18, 19, 23, 26, 27, 28, 29, 31, 32, 33, 34, 35, 37, 38, 39, 40, 43, 44, 45, 46, 47, 49, 50, 51, 52, 53, 54, 57, 58, 59, 60, 61, 62, 69, 70, 71, 72, 73, 75, 77, 78, 79, 80, 81, 82, 83, 87, 89, 90, 91, 92, 95, 97, 99, 100, 101, 103, 105, 106, 107, 108, 109, 110, 111,

113, 116, 117, 118, 119, 121, 122, 126, 128, 129, 130, 131, 136, 137, 138, 139, 140, 141, 145
cell culture, 26, 40, 137
cell cycle, 7, 57, 81, 97, 105, 113
cell death, 5, 40, 61, 100, 105, 111, 113, 138
cell fate, 7
cell invasion, 35, 50, 113, 145
cell killing, 59, 60, 92
cell line, 7, 11, 16, 35, 57, 62, 80, 92, 103, 106, 107, 108, 109, 116, 117, 122, 139
cell membranes, 117
cell metabolism, 28
cell signaling, 105
cell surface, 4, 13, 27, 33, 46, 78, 117, 140
central nervous system, 38, 47, 130
central retinal vein occlusion, 125, 126, 133
chemokines, 26, 32
chemotaxis, 19
chemotherapeutic agent, ix, 11, 55, 56, 59, 60, 61, 67, 112
chemotherapy, viii, ix, 1, 5, 10, 11, 12, 14, 17, 20, 24, 27, 55, 56, 57, 58, 59, 60, 61, 62, 63, 64, 65, 66, 67, 68, 69, 70, 71, 72, 73, 74, 75, 79, 105, 118, 119
childhood, 17, 104
children, vii, ix, 1, 15, 55, 57, 64, 72
China, 106
cholangiocarcinoma, 17
chondrocyte, 34
choroid, 50
chromosome, 27
CIA, 24, 26, 36, 37, 38
circulation, vii, 13, 61, 101, 121
cirrhosis, vii, 1, 3, 10, 13, 18
classes, 100, 102, 110
cleavage, 34, 37, 117
clients, 102
clinical oncology, 68
clinical trials, 7, 8, 11, 12, 13, 14, 21, 69, 79, 94, 101, 102, 110, 119, 124
cloning, 116
clusters, 139
coagulation, x, 17, 99, 100, 105, 123
coding, 129, 138
coherence, x, 115, 122, 124, 133
cohort, 73
collaboration, 50
collagen, 7, 18, 24, 26, 32, 33, 34, 36, 37, 38, 39, 41, 47, 48, 50, 51, 52, 53, 54, 139
colon, x, 27, 62, 64, 73, 77, 79, 80, 83, 84, 86, 91, 94, 95, 96, 107, 108, 109, 119
colon cancer, x, 64, 77, 79, 80, 83, 84, 86, 91, 94, 95, 96, 107, 109
colorectal adenocarcinoma, 95
colorectal cancer, viii, 11, 17, 20, 24, 27, 37, 40, 61, 64, 71, 73, 111, 118, 119, 123, 131
combination therapy, 94
commitment, 29
communication, 121
complement, 78
compliance, 80
complications, xi, 39, 119, 125, 135, 136, 137, 141, 142
components, 28, 32, 59, 104, 113
composition, 26
compounds, 11, 38, 100, 102, 103, 105, 110
computed tomography, 8
concentration, 34, 41, 59, 61, 69, 81, 83, 84, 103, 118, 122
congestive heart failure, 119, 120
connective tissue, 32, 47
consensus, 40, 104
constipation, 8, 94
construction, 27
control, x, xi, 5, 7, 8, 10, 11, 29, 40, 80, 82, 84, 85, 86, 87, 88, 91, 107, 115, 120, 121, 123, 124, 135, 136, 141
control group, 5, 7, 10, 11, 84, 85, 91, 124
controlled trials, ix, x, 56, 68, 73, 116, 128
conversion, 137
correlation, 5, 7, 12, 20, 28, 41, 66, 74, 88, 97, 130
cost saving, 127
coverage, 63, 136, 138, 144
cryotherapy, viii, 1
CSF, 9
culture, 33, 35, 46, 57, 81, 108, 136
cycles, ix, 55, 57, 58, 60, 61
cycling, x, 57, 94, 99, 100, 105
cyclooxygenase, 17, 18, 65, 66, 74, 109
cyclophosphamide, 60, 66, 69, 71, 72, 74, 75
cystoid macular edema, 126, 127, 133
cytochrome, 106, 111
cytokines, 4, 26, 27, 28, 29, 30, 32, 39, 42, 43
cytoplasmic tail, 6
cytotoxicity, 61, 67, 78, 81, 83, 107

D

damage, 60, 61, 62, 100

death, 4, 39, 60, 67, 92, 113
deaths, vii, 1
defects, 4, 34
defense, 100
deficiency, 50, 137
definition, 126
degradation, 2, 4, 26, 31, 33, 35, 46, 102, 103
delivery, 2, 51, 59, 61, 79, 81, 95, 118
denaturation, 33
Denmark, 82
density, 5, 6, 10, 12, 20, 29, 37, 59, 82, 87, 105, 130
deposits, 93
depression, 88, 94
derivatives, 54, 103, 110, 111, 137
destruction, 25, 26, 29, 32, 36, 37, 43, 51, 58, 60, 69, 105
destructive process, 32
detachment, 124, 140
detection, viii, 23, 44
diabetes, xi, 126, 135, 136, 137, 138, 139, 141, 144, 145
diabetic patients, xi, 135, 136, 137
diabetic retinopathy, v, x, xi, 115, 117, 125, 126, 127, 130, 132, 135, 136, 137, 138, 139, 140, 141, 142, 143, 144
diarrhea, 85, 88
diet, 107
differentiation, 103, 116
direct action, x, 77, 79
direct measure, 12
disease activity, 41, 51
disease progression, 10, 35, 123
distribution, 81, 85, 93, 94, 95, 97, 118, 132
diversity, 57
division, 28
DNA, 35, 48, 104, 108, 110, 122, 138
DNA breakage, 110
docetaxel, 60, 73
domain, 6, 7, 31, 32, 33, 34, 37, 48, 101, 102, 105, 139, 145
dosage, 82
dosing, ix, 10, 14, 56, 59, 61, 62, 63, 64, 67, 68, 70, 72, 75, 94
down-regulation, 19, 70, 140, 143
drinking water, 62, 71
drug delivery, 96
drug discovery, 26
drug resistance, ix, 17, 56, 60, 62, 79
drug therapy, 68
drugs, ix, 19, 24, 25, 37, 42, 56, 58, 62, 63, 64, 65, 68, 69, 70, 78, 105, 118
duration, 8, 9, 11, 28, 59, 65, 66
dwarfism, 47
dysplasia, 119

E

edema, 124, 126, 127, 133, 140
egg, vii
elastin, 32
elderly, 64, 73
electrons, 93, 99
electroretinography, 122
emboli, 93
embolus, 5
embryo, 26, 29, 107
emitters, 93
EMMPRIN, 33
emphysema, 124
encoding, 11
endocrine, 68, 117
endometrial carcinoma, 130
endothelial cells, viii, ix, 2, 4, 6, 13, 15, 18, 19, 21, 26, 28, 29, 31, 32, 34, 35, 38, 43, 46, 48, 49, 50, 55, 56, 58, 69, 70, 72, 75, 101, 103, 104, 107, 108, 109, 112, 116, 117, 122, 128, 129, 132, 139, 141, 142, 143, 144, 145
endothelium, 2, 4, 12, 13, 69
energy, 93, 94
enlargement, 34
enrollment, 127
environment, 4
enzymes, 26, 30, 31, 32, 33, 100, 104
epithelial cells, 129, 137
epithelium, xi, 69, 117, 122, 135, 136, 138, 141, 142, 143, 144, 145, 146
erosion, 25, 36
erythrocytes, 80
erythropoietin, 31
estrogen, 54, 75, 76, 94, 97
etanercept, 26, 40, 42
ethanol, 10
evening, 9
evidence, ix, 25, 35, 36, 55, 56, 59, 61, 67, 83, 105, 128, 137, 139, 140
evoked potential, 121
evolution, 25, 47
exclusion, 81, 119, 124
exons, 27, 117

exposure, 48, 58, 59, 81, 83, 122
expression, viii, 4, 5, 6, 7, 8, 10, 13, 16, 17, 18, 19, 21, 23, 26, 27, 28, 29, 33, 34, 35, 37, 41, 42, 43, 44, 48, 49, 50, 51, 52, 53, 68, 76, 95, 101, 103, 118, 129, 130, 136, 137, 138, 139, 140, 143, 144, 145
extracellular matrix, 3, 4, 5, 26, 27, 32, 33, 34, 35, 48, 103, 107, 110
extravasation, 52, 93

F

failure, 127
family, 3, 5, 6, 27, 30, 32, 36, 37, 43, 52, 102, 105, 106, 111, 117, 129
fatigue, 8
FDA, 78, 118, 123, 127
FDA approval, 118, 123, 127
feet, 28
females, vii, 118
fetus, vii
fever, 69
fibers, 145
fibrin, 26, 34, 48
fibrinolysis, 17
fibroblast growth factor, 10, 16, 19, 24, 27, 42, 44, 61, 66
fibroblasts, 27, 29, 30, 31, 32, 35, 37, 38, 42, 43, 44, 45, 46, 50, 112
fibrosarcoma, 50
fibrosis, 33, 36
filtration, 80
Finland, 25
fluid, 25, 30, 32, 37, 124, 128, 130, 138, 143
focusing, xi, 32, 36, 135, 136
follicle, 57
food, 101
fractures, 26
fructose, 142
fuel, 34
funduscopic examination, 121
fungi, 100

G

gadolinium, 12
ganglion, 117, 122
gastric mucosa, 80
gastrointestinal tract, 117, 130
gelatinase A, 32, 48, 49
gene, 3, 8, 16, 17, 18, 27, 35, 43, 50, 52, 53, 117, 129, 130, 137, 139, 140
gene expression, 16, 50, 130, 139
gene therapy, 18, 53
gene transfer, 43, 52, 53, 140
generation, 28, 31, 43, 107, 117, 136, 137, 138, 143, 144
genes, vii, 6, 19, 31, 45, 50, 70, 103, 109
genetic factors, 25
genetics, 111
glaucoma, 126, 132, 133
glioblastoma, 66, 75, 109, 117
glioblastoma multiforme, 117
glioma, 71, 74, 103, 113
glucose, xi, 30, 31, 44, 135, 136, 141, 142
GLUT, 31
glutamate, 100
glutathione, 100, 107
glycolysis, 31
glycoproteins, 27
glycosylation, 142, 143
gold, 25
Greece, 1
groups, 10, 28, 32, 33, 38, 62, 81, 82, 83, 88, 119, 127, 137
growth, vii, viii, ix, x, 2, 3, 5, 6, 7, 11, 12, 15, 16, 17, 18, 19, 20, 21, 24, 27, 28, 29, 30, 33, 34, 36, 37, 38, 40, 41, 42, 43, 44, 45, 48, 49, 50, 51, 52, 53, 54, 56, 57, 58, 61, 62, 63, 66, 68, 70, 71, 74, 75, 76, 77, 79, 80, 85, 86, 91, 92, 93, 95, 97, 98, 100, 101, 102, 105, 107, 108, 109, 112, 115, 117, 118, 119, 128, 129, 130, 131, 132, 136, 137, 138, 139, 140, 141, 142, 143, 144, 145, 146
growth factor, vii, x, 2, 3, 5, 6, 15, 16, 17, 18, 20, 24, 27, 28, 29, 30, 36, 37, 40, 41, 42, 43, 44, 45, 48, 49, 50, 51, 52, 53, 58, 63, 66, 68, 70, 71, 74, 75, 76, 97, 109, 112, 115, 117, 128, 129, 130, 131, 132, 137, 139, 142, 143, 144, 145, 146
growth rate, 98, 101
gut, 57, 63

H

half-life, 30
HE, 50, 68, 72, 97
healing, 26, 119
health, vii, 141
heat, 33, 112, 113
heat shock protein, 112, 113

hemorrhage, 119, 120, 125
hepatic fibrosis, 7
hepatic stellate cells, 4, 50
hepatitis, 7, 10, 11, 13, 18, 20
hepatitis a, 13
hepatocarcinogenesis, 10, 106
hepatocellular carcinoma, vii, 1, 3, 10, 12, 14, 15, 16, 17, 18, 19, 20, 21
hepatocytes, 3, 6
hepatoma, 4, 7, 20, 107, 111
herbal medicine, 106
heterogeneity, 68, 75
histology, 71
HIV, 37
homeostasis, 52, 136
homogeneity, 94
hormone, 65, 66, 74
host, vii, 15, 50, 57
human immunodeficiency virus, 94
human lung fibroblasts, 47
human subjects, x, 43, 115, 122
hybridoma, 78, 80
hydrogen, 100, 110
hydrogen peroxide, 100, 110
hydroxyl, 100
hyperglycemia, xi, 135, 136
hyperplasia, 16, 37
hypertension, 11, 119, 123, 124, 126
hyperthermia, 63, 72, 79, 95
hypoglycemia, xi, 135
hypothesis, x, 3, 26, 31, 39, 59, 61, 77, 79
hypoxia, 3, 4, 15, 24, 28, 29, 30, 31, 32, 35, 42, 45, 50, 51, 52, 72, 101, 108, 113, 117, 130
hypoxia-inducible factor, 24, 30, 45, 51, 101, 113
hypoxic cells, 45

I

ICAM, 19, 42, 139, 144
identification, viii, ix, 23, 25, 56, 67, 116, 129
idiopathic, 41
IFN, 9, 10, 60
IL-6, 26, 28
IL-8, 26
images, 12
immune response, 54
immunodeficiency, 53, 62, 71
immunoglobulin, 26, 122
immunohistochemistry, 121
immunomodulatory, 9
immunoreactivity, 12, 80
immunotherapy, 63, 68, 72
in vitro, ix, 6, 7, 8, 13, 14, 17, 18, 19, 27, 28, 35, 48, 49, 50, 54, 56, 57, 60, 67, 70, 72, 74, 80, 97, 102, 105, 106, 109, 111, 113, 117, 122, 130, 132, 133, 137, 138, 142, 143
incidence, viii, 6, 24, 36, 38, 54
independence, 93
inducer, 30, 33, 116
induction, 7, 12, 26, 28, 30, 31, 38, 45, 63, 71, 92, 107, 111, 129
infancy, 70
infants, 19, 64, 72
infection, 7, 26, 39, 85, 88, 94
infinite, 80
inflammation, 13, 25, 28, 36, 37, 39, 45, 139
inflammatory arthritis, 41, 44, 53
inflammatory cells, 121
inflammatory disease, 25, 39
infliximab, 26, 28, 42
informed consent, 128
inhibition, viii, 6, 8, 10, 17, 19, 23, 24, 26, 36, 38, 50, 51, 52, 53, 63, 69, 72, 85, 86, 92, 102, 103, 104, 106, 107, 109, 113, 119, 122, 137, 145
inhibitor, 5, 6, 7, 10, 11, 12, 17, 18, 19, 20, 24, 34, 36, 37, 38, 43, 46, 47, 49, 50, 51, 53, 60, 62, 63, 65, 66, 73, 74, 95, 97, 103, 105, 107, 113, 137, 139, 140, 141, 145
initiation, 120
injections, x, 105, 116, 121, 122, 123, 124, 125, 126
injury, vii, 44, 93, 136, 141, 143
inoculation, 80, 81, 82, 83, 89, 90, 91
instability, 102
insulin, 141
integration, 79
integrin, 2, 8, 28, 34, 49, 103
integrity, 138
intensity, 12, 13, 58, 59, 72
interaction, 2, 7, 26, 33, 95, 137, 138, 139
interactions, 25, 138, 139, 141
intercellular adhesion molecule, 144, 145
interest, 37, 39, 63, 139
interface, 25, 31
interference, 67
interferon, 10, 13, 19, 20, 21, 60, 66, 70, 74
interferon (IFN), 60
internalization, 80
interval, 58, 61, 123
intervention, 25, 95, 97
intraocular, 126, 132, 140

intraocular pressure, 126
intravenously, x, 81, 82, 115
introns, 27, 117
investment, 2
iodine, 97
iris, 116, 125, 126, 132, 133
iron, 31, 100, 105
irradiation, 60, 81, 84, 92
ischemia, 32, 137, 140, 141
IV collagenase, 47

J

Japan, 55, 64, 77, 80, 81, 135, 140
joint destruction, 28, 31, 32, 41, 51
joint swelling, 25
joints, 25, 30, 36, 37, 44, 45, 52
justification, 128
juvenile rheumatoid arthritis, 41

K

Kaposi sarcoma, 66
ketones, 137
kidney, 113, 117, 130
killing, 59, 93
kinase activity, 27
kinetic model, 57
kinetics, 57, 63, 75, 98
knowledge, viii, 2, 56
kringle domain, 37

L

labeling, 105, 121
lactation, 120
laminar, 127
laminin-5, 47
leakage, 116, 125, 126
Lebanon, 99
leptin, 143
lesions, x, 10, 77, 79, 92, 93, 106, 113, 123, 124, 136
leukemia, 57, 62, 78, 107, 111
leukopenia, 66
LFA, 19
life span, 106
ligand, 2, 4, 15, 60, 118, 129
linkage, 102
liver, vii, x, 1, 3, 5, 7, 15, 16, 18, 20, 66, 71, 77, 79, 80, 82, 88, 89, 91, 92, 94, 95, 96, 106, 113
liver cancer, 20
liver cells, 7
liver cirrhosis, 5
liver disease, 16
liver metastases, 82, 88, 89, 91, 95, 96
localization, 15
lumen, 27
lung cancer, x, 37, 69, 72, 73, 78, 109, 115, 118, 119, 130, 131
lymphocytes, 27
lymphoma, 61, 70, 71, 73, 78, 95, 109

M

macromolecules, 100
macrophages, 4, 25, 27, 29, 31, 32, 45, 47, 101, 109
macular degeneration, x, 115, 116, 129, 131, 132
magnetic resonance, 12
magnetic resonance imaging, 12
males, 118
malignancy, 6, 17
malignant melanoma, 74, 107
management, viii, 14, 24, 26, 38, 60, 68, 132
mandible, 70
manipulation, 92
mantle, 95
marrow, x, 21, 29, 31, 51, 70, 77, 94
mass, 30, 32
matrix, 2, 4, 5, 6, 10, 16, 25, 26, 28, 33, 35, 42, 43, 46, 47, 48, 49, 50, 51, 103, 109, 140, 145
matrix metalloproteinase, 2, 10, 16, 42, 43, 46, 47, 48, 49, 50, 51, 103, 109, 140
maturation, 30, 38, 69
MCP, 109, 139
MCP-1, 109, 139
measles, 106
measurement, 12, 13, 46
media, 116, 137
median, 8, 9, 10, 65, 66, 118
medication, 118, 123, 127
melanoma, 38, 62, 63, 66, 72, 74, 107
membranes, 37, 117, 129
menadione, 110
mesentery, 105
mesoderm, 29
metabolism, 28, 31, 98, 100, 104, 105, 110
metabolites, 97, 113
metalloproteinase, 3, 15, 24, 26, 40, 46, 47, 49, 50

metastasis, x, 3, 5, 10, 15, 17, 19, 49, 59, 71, 77, 79, 80, 82, 88, 92, 93, 94, 95, 96, 97, 99, 100, 129
metastatic cancer, 121
metastatic disease, ix, 55, 57
methanol, 81
Miami, 123
mice, 8, 10, 11, 16, 19, 20, 33, 34, 36, 37, 38, 47, 48, 49, 51, 53, 54, 60, 61, 62, 63, 68, 71, 80, 81, 82, 83, 85, 87, 88, 89, 90, 91, 94, 95, 96, 97, 105, 106, 109, 117, 130, 136, 139
microscopy, 82, 121
migration, x, 2, 5, 6, 7, 31, 32, 34, 44, 47, 48, 51, 99, 101, 103, 108, 109, 113, 137, 140, 141
Ministry of Education, 140
minority, vii, 1
mitogen, 17, 128, 129, 137, 138
mitosis, 62
MMP, 2, 3, 4, 6, 10, 13, 14, 15, 24, 26, 32, 33, 34, 35, 46, 47, 48, 50, 109, 140
MMP-2, 2, 4, 15, 32, 33, 34, 35, 46, 50, 140
MMP-3, 5, 32, 35
MMP-9, 2, 5, 10, 15, 32, 34, 35, 46
MMPs, 2, 4, 34, 46, 47, 48, 140
mode, 95, 102, 107, 110, 128
modeling, 60
models, v, viii, ix, 11, 12, 24, 25, 26, 31, 36, 37, 38, 51, 56, 57, 61, 77, 80, 93, 97, 101, 102, 103, 137
molecular oxygen, 100
molecular weight, 28
molecules, vii, viii, 2, 10, 13, 19, 23, 25, 26, 29, 31, 35, 36, 37, 42, 60, 99
monitoring, ix, 56, 67, 68, 119, 127
monoclonal antibody, x, 11, 39, 62, 71, 78, 79, 80, 94, 95, 96, 97, 115, 125, 131
monocyte chemoattractant protein, 139, 144
Moon, 16, 48
morbidity, 127
morning, 9
morphogenesis, 51
morphology, 30, 40, 54
mortality, 25
movement, 30
MRI, 12, 20, 71
mRNA, 4, 27, 28, 29, 35, 44, 101, 107, 109, 117, 138, 139, 143
mucosa, 6, 57, 63, 80
mutant, 102
mutation, 130
myeloid cells, 31
myocardial infarction, 119, 123, 124
myopia, 125, 132

N

naphthalene, 105
nausea, 94
necrosis, viii, 23, 24, 25, 39, 42, 43, 92, 143
needs, 26
negativity, 7
neoangiogenesis, 6, 92
neovascularization, x, 2, 3, 7, 8, 10, 40, 43, 48, 70, 72, 79, 93, 97, 101, 106, 107, 115, 116, 125, 126, 132, 136, 137, 138, 140, 141, 144, 145
nervous system, 111
network, vii, 20, 25, 29, 105, 138, 144
neuroblastoma, 34, 49, 62, 64, 72, 105
neurotoxicity, 9, 38
neutropenia, 64, 66, 69, 94
neutrophils, 25, 29, 32, 42, 138, 143
New Zealand, 121
nitric oxide, 6, 18, 126, 144
nitric oxide synthase, 18, 144
nodules, 6, 10, 12, 20, 25, 32, 80, 82, 94
non-Hodgkin's lymphoma, 64
non-steroidal anti-inflammatory drugs, 25
normal development, 47
nuclear magnetic resonance, 44, 98
nucleophiles, 100
nucleus, 31, 45
nutrients, vii, viii, 2, 24, 27, 30, 39, 118

O

observations, viii, 24, 30, 37, 128, 137, 138, 139, 140
occlusion, 117, 136
ocular diseases, x, xi, 115, 116, 128
oedema, 25
oil, 107
optic nerve, 116
optimization, 19, 70
organ, 26, 46, 93
organism, 99
organization, ix, 55, 145
orthostatic hypotension, 94
ossification, 33, 48
osteoarthritis, 40, 43, 45, 46
osteocyte, 47
osteoporosis, 112

ovarian cancer, 64, 73, 74, 79, 96
ovaries, 130
ovulation, vii
oxalate, 82
oxidative stress, 100, 137, 139, 144
oxygen, vii, viii, 24, 27, 30, 31, 39, 44, 99, 101
oxygen consumption, 30

P

p53, 102, 103, 106, 107, 111, 130
paclitaxel, 59, 62, 68, 73, 95, 118, 131
pain, 25
palliative, 64, 65, 75
pancreatic cancer, 97, 107
pannus formation, 36, 37, 53
paradigm shift, 56
parenchyma, 3
parental control, 107
particles, 93
passive, 53
pathogenesis, 6, 25, 30, 116, 137, 138, 140, 141, 142
pathophysiology, viii, 2, 112, 142
pathways, 6, 7, 11, 14, 35, 67, 102, 105, 107
peptides, 18
perforation, 119, 120
perfusion, 12, 20, 30, 94
peripheral blood, 29, 38, 43, 83, 87
peripheral neuropathy, 8, 94
peritoneal carcinomatosis, 96
permeability, 12, 41, 44, 111, 116, 117, 126, 128, 129, 130, 138, 140, 145
peroxide, 100, 136, 141
perspective, 46
pH, 33, 118
pharmacokinetics, 59, 95, 132
pharmacology, 14, 111
phenotype, 5, 33, 49, 50, 60, 75
phosphorylation, 37, 101, 103, 110, 112
photosynthesis, x, 99, 100
pilot study, 21, 41, 66, 74
pioglitazone, 65, 66, 74
placebo, 39
placenta, vii, 129
planning, 127
plants, 100, 106
plasma, 52, 105, 136
plasma proteins, 105
plasminogen, 2, 5, 17, 37, 53, 103, 140, 145, 146
plasticity, 144
platelet aggregation, 6
platelet count, 3, 16, 66, 82, 88
platelets, 27, 87
platinum, 74
PM, 142
polypeptide, 44
poor, 13, 15, 18, 130
population, 2, 39, 57, 62, 66, 93, 124
portal vein, 3, 4, 5
positive correlation, 3, 7
positive feedback, 138
power, 82
precursor cells, 2, 21, 29, 43
predictors, 10
prednisone, 64
pregnancy, vii, 38, 120
prematurity, 137
preservative, 118
pressure, 30, 93, 118, 126
prevention, 107, 142
primary tumor, 2, 6, 20, 93
primate, 125, 129, 132
principle, 27, 56
probability, 58
prodrugs, 104
production, vii, 6, 26, 33, 39, 41, 42, 45, 47, 52, 53, 65, 80, 100, 103, 117, 126, 128, 137, 138, 144
prognosis, 6, 18, 20, 21, 25, 130
proliferation, vii, x, 2, 5, 7, 10, 11, 12, 16, 25, 28, 29, 32, 34, 38, 40, 49, 53, 60, 70, 72, 81, 83, 84, 97, 99, 101, 105, 106, 107, 115, 116, 117, 122, 123, 137, 140, 141, 145
promoter, 35, 52
prostaglandins, 6
prostate, 47, 63, 65, 66, 71, 74, 79, 103, 109
prostate cancer, 63, 66, 74, 79, 109
prostate carcinoma, 47, 74, 109
prostate specific antigen, 66
protease inhibitors, 137, 138
protective role, 139, 140
protein kinase C, 136
proteinase, 7, 47
proteins, vii, 5, 33, 44, 47, 76, 100, 101, 102, 103, 105, 112, 113, 117, 137, 142
proteinuria, 11, 119
proteoglycans, 139
proteolysis, 140
proteolytic enzyme, 26, 32
protocol, 63, 64, 85, 127
pruning, 116

pseudopodia, 34
psoriasis, 116
psoriatic arthritis, 28, 30

Q

quality of life, ix, 55, 57
quinone, 99, 100, 102, 104, 108, 109, 110, 112
quinones, x, 99, 100, 101, 102, 105, 110

R

radiation, 60, 79, 82, 84, 91, 92, 93, 94, 95, 97, 126
radiation therapy, 94
radiosensitization, 81
radiotherapy, 60, 70, 73, 96
range, 27, 38, 49, 60, 94, 118, 125
rash, 8
reactive arthritis, 40
reactive oxygen, 100, 111, 136, 143, 144
reactivity, 80
reasoning, 59
receptors, 2, 3, 4, 5, 8, 11, 13, 15, 26, 27, 28, 29, 36, 37, 40, 43, 52, 60, 75, 101, 103, 117, 130, 142, 144, 145
recovery, ix, 8, 55, 57, 61, 88, 94
rectum, 27
recurrence, vii, 1, 5, 10, 20
reduction, 11, 12, 28, 30, 37, 38, 53, 61, 63, 65, 67, 87, 99, 124, 126
reflection, 91
regression, ix, 12, 16, 20, 55, 57, 62, 63, 71, 87, 126
regrowth, 57, 58, 61
regulation, x, 4, 6, 10, 16, 18, 19, 32, 33, 35, 42, 45, 47, 48, 50, 51, 52, 75, 115, 129, 139, 140, 143, 145, 146
regulations, 80
regulators, 40, 67, 75
relationship, 11
relationships, 68, 130
remission, 20, 63
remodelling, 5, 32, 144
renal cell carcinoma, 51
renal dysfunction, 123
renin, 136, 145
repair, 36, 52, 62, 116
replication, 8, 104
repressor, 17
reproduction, 27, 36
resection, vii, 1, 3, 10, 19, 20, 70, 73
residual disease, 61, 79
residues, 30
resistance, 59, 62, 68, 104
resolution, ix, 56, 68, 126, 127
resources, 128
respiration, x, 99
respiratory, 25, 100
responsiveness, viii, 24
retardation, 138
retention, 31, 81, 85, 92, 95
retina, x, 115, 116, 121, 124, 128, 144
retinal detachment, 125, 140, 142, 145
retinal ischemia, 132
retinoblastoma, 138
retinopathy, xi, 125, 126, 135, 136, 137, 138, 139, 140, 141, 144, 145
rheumatoid arthritis, viii, 23, 24, 39, 40, 41, 42, 43, 44, 45, 46, 47, 51, 53, 116
rice, 32
risk, ix, xi, 11, 19, 25, 26, 39, 40, 56, 62, 73, 119, 127, 128, 135, 136, 140, 141
risk profile, 40
rituximab, 66
RNA, 5, 10, 129
rodents, 54, 80, 82

S

sacrifice, 57, 81
safety, 26, 121, 122, 127, 131, 132
sample, 33, 82
savings, 127
scatter, 50, 127
scheduling, 20, 59, 61, 68, 69, 73, 92
sclera, 25
second generation, 9
secrete, 32, 42, 47, 110, 117, 128
secretion, 30, 35, 43, 52, 101, 103, 108, 109
seed, 111
sensitivity, 13, 57, 60, 79, 94, 101
series, vii, x, 10, 38, 66, 92, 109, 115, 124, 125
serine, 5, 137, 138
serology, 11
serum, 3, 5, 7, 10, 15, 16, 18, 28, 31, 32, 41, 42, 44, 46, 49, 65, 66, 123, 130
severity, viii, 23, 26, 36, 37, 38, 51, 53, 54, 62, 139
sex, 25
sex hormones, 25
shock, 112

side effects, xi, 8, 119, 125, 135, 136
signal transduction, 3, 29
signaling pathway, x, 5, 16, 99, 103
signalling, 28, 29, 36, 40
signals, vii, 4, 6, 28, 138
signs, viii, 23, 36, 85, 88, 121
sites, 17, 29, 33, 34, 47, 101, 129, 140
skin, 8, 71, 94
slit lamp, 126
slit lamp examination, 126
smoking, 124
smooth muscle, 2, 4, 27
smooth muscle cells, 2, 4
sodium, 118, 138, 143
soft tissue sarcomas, 104
solid tumors, ix, x, 9, 55, 57, 64, 77, 78, 79, 93, 94, 104
solubility, 33
somnolence, 94
species, 33, 100, 111, 129, 136, 143, 144
specificity, 13, 94, 145
spectroscopy, 44
spectrum, 97
spinal cord, 37
spleen, 80
sprouting, 5
squamous cell, 130
squamous cell carcinoma, 130
stability, 101, 102
stabilization, 7, 9, 18, 47
stages, 7, 26, 93
stem cells, 70
sterile, 118
stimulus, viii, 3, 24, 26, 27, 31
strain, 31, 108
strategies, v, xi, 1, 11, 14, 79, 94, 97, 135, 136
stress, 112, 113, 137, 145
stroke, 119, 123
stroma, 3, 58, 69
subacute, 5
substitution, 137, 140
substrates, 33
sugar, 104
Sun, 15, 16, 105, 113
supply, viii, 24, 28, 30, 92, 101
suppression, x, 54, 69, 77, 79, 91, 92, 93, 106, 113, 119, 143
surface area, 12
survival, vii, ix, 1, 7, 8, 9, 10, 11, 16, 18, 27, 30, 31, 49, 55, 57, 59, 61, 63, 64, 66, 70, 72, 75, 79, 81, 83, 84, 91, 96, 102, 109, 118, 130, 137, 138
survival rate, 8, 10, 66
survivors, 81
Sweden, 80
swelling, 25, 31, 36, 37
switching, 16
symptoms, viii, x, 23, 36, 77, 79
syndrome, 117
synovial fluid, 25, 26, 28, 30, 32, 36, 37, 40, 41, 42, 44, 46, 52
synovial membrane, 31, 32, 37, 40, 46
synovial tissue, viii, 24, 25, 29, 32, 38, 41, 43, 44, 53
synovitis, 25, 28, 31, 35, 36, 42
synthesis, 6, 34, 48, 109, 122
systemic sclerosis, 28
systems, 136
systolic blood pressure, 123

T

T cell, 25, 139, 145
T lymphocytes, 63
tamoxifen, 73, 75
targets, viii, ix, 14, 23, 25, 30, 31, 35, 51, 55, 58, 59, 69, 70, 101, 104
taxane, 64
taxanes, ix, 56
technetium, 95
technology, 70, 78
tendon, 25, 29, 32, 35, 39, 43, 45, 46, 47
tendons, 25, 29, 32
tenosynovitis, 25
tension, 30, 44
TGF, 2, 6, 24, 27, 42, 52
theory, vii, 68, 93
therapeutic agents, 17, 97
therapeutic targets, 16, 37
therapeutics, viii, 2, 13, 112
therapy, v, viii, ix, x, xi, 6, 10, 14, 18, 19, 20, 21, 23, 24, 27, 37, 39, 40, 41, 52, 53, 56, 57, 59, 60, 62, 63, 64, 65, 67, 68, 69, 70, 71, 73, 74, 77, 79, 80, 81, 82, 83, 88, 90, 91, 92, 93, 94, 95, 96, 97, 104, 110, 111, 118, 123, 124, 125, 126, 127, 131, 132, 135, 136
thinking, 59
threshold, 59
thrombosis, 3, 119, 120
thyroid, 130

thyroid cancer, 130
time, 8, 10, 26, 27, 36, 42, 57, 58, 59, 60, 62, 64, 66, 68, 69, 103, 118, 119, 121, 123, 126, 128
TIMP, 24, 32, 33, 34, 35, 46, 47, 49, 50
TIMP-1, 33, 34, 35, 49
tirapazamine, 72
tissue, vii, viii, ix, 5, 12, 16, 21, 24, 25, 26, 28, 29, 30, 31, 32, 33, 35, 43, 46, 47, 49, 50, 52, 55, 58, 59, 63, 66, 71, 74, 104, 105, 116, 132, 142
TNF, viii, 8, 19, 23, 26, 37, 39, 40, 42, 51, 54, 101, 107, 109, 143
TNF-alpha, 19, 42, 51, 54
TNF-α, 101, 107, 109, 143
topology, 104
toxic effect, 66, 72
toxicity, ix, x, 20, 55, 56, 57, 58, 59, 62, 63, 65, 66, 67, 68, 71, 72, 77, 79, 82, 83, 92, 94, 95, 101, 102, 104, 111, 121, 122, 123, 131, 138, 143
toxin, 104
transcription, 31, 33, 45, 101, 103, 104, 109, 137
transcription factors, 45
transcripts, 27
transducer, 6
transduction, 45
transformation, 102, 113
transforming growth factor, 6, 24, 27, 42
transient ischemic attack, 123, 124
transition, 111
translocation, 45
transplantation, vii, 1
transport, 79, 100
trauma, 46
trial, 8, 14, 19, 21, 26, 39, 40, 59, 61, 68, 73, 75, 95, 118, 123, 127, 131
triggers, 15, 49, 111
trypsin, 80
tuberculosis, 26
tumor, vii, viii, ix, x, 1, 2, 3, 4, 5, 7, 8, 10, 11, 12, 13, 14, 15, 17, 18, 20, 21, 39, 40, 41, 42, 43, 46, 47, 48, 49, 50, 51, 55, 56, 57, 58, 59, 60, 61, 62, 63, 67, 68, 69, 70, 71, 72, 77, 78, 79, 81, 82, 85, 86, 87, 91, 92, 93, 97, 99, 100, 101, 103, 104, 105, 106, 107, 109, 110, 111, 112, 113, 116, 117, 118, 140, 145
tumor cells, vii, ix, x, 2, 13, 55, 56, 57, 58, 59, 60, 61, 62, 67, 77, 78, 79, 82, 92, 93, 94, 101, 103, 104, 117
tumor growth, x, 11, 12, 17, 18, 48, 49, 50, 57, 59, 61, 62, 69, 85, 86, 91, 92, 99, 100, 106, 107, 109, 140, 145
tumor invasion, 5, 15, 46, 47
tumor metastasis, 110
tumor necrosis factor, 8, 39, 41, 42, 43, 46
tumor progression, 5, 21, 49, 112
tumors, vii, ix, 1, 2, 5, 6, 7, 12, 13, 16, 57, 59, 60, 62, 63, 64, 66, 69, 70, 71, 74, 77, 79, 81, 82, 86, 87, 88, 92, 93, 96, 97, 98, 105, 107, 111, 117
tumour growth, viii, 24, 30, 34, 36, 38, 39, 54, 98, 131
tumours, 27, 34, 37, 40
turnover, 28, 33, 47, 52
type 2 diabetes, 141
tyrosine, 3, 6, 11, 17, 18, 20, 27, 36, 37, 43, 48, 51, 52, 63, 103, 105, 129

U

UK, 23, 141
United Kingdom, 23
United States, 78, 113
unstable angina, 119
urine, 37, 94
urokinase, 5, 17, 103
uterus, vii

V

validation, v, 77
variable, 10, 13, 14, 59
variables, 13, 105
variance, 7
variation, 60
vascular diseases, x, 116
vascular endothelial growth factor (VEGF), viii, 2, 15, 24, 27, 42, 52, 53, 60, 79, 101, 116, 129, 130, 137, 138
vascular wall, 137
vasculature, viii, 2, 4, 11, 12, 14, 15, 23, 24, 26, 51, 61, 62, 93, 101, 103, 104, 116, 118, 142
vasculitis, 25
VCAM, 42
vector, 3, 11, 34, 37, 53
VEGF expression, 3, 6, 28, 29, 30, 31, 37, 138
vein, 48, 82, 117
vessels, vii, 2, 12, 25, 26, 28, 30, 34, 37, 60, 93, 100, 109, 125, 136, 143
vision, x, 78, 115, 124, 125, 126, 128
visual acuity, 124, 127
vitamin K, 105, 112, 113

vitreous hemorrhage, 125, 132

W

water, 62
weapons, 128
weight changes, 88
weight loss, 62, 85, 88, 94
wild type, 34
withdrawal, 65
women, 73
wound healing, 11, 63, 119
writing, 36

X

xenografts, 10, 11, 38, 62, 63, 71, 72, 80, 84, 85, 86, 92, 95, 96, 105, 109

Z

zinc, 32